Winning the Battle of the Airfields

Seventy Years of RAND Analysis on Air Base Defense and Attack

ALAN J. VICK AND MARK ASHBY

Prepared for the Department of the Air Force
Approved for public release; distribution unlimited

 PROJECT AIR FORCE

For more information on this publication, visit www.rand.org/t/RRA793-1

Library of Congress Cataloging-in-Publication Data is available for this publication.
ISBN: 978-1-9774-0661-3

Published by the RAND Corporation, Santa Monica, Calif.
© 2021 RAND Corporation
RAND® is a registered trademark.

Cover: Tech. Sgt. Natasha Stannard/U.S. Air Force

Support RAND
Make a tax-deductible charitable contribution at
www.rand.org/giving/contribute

www.rand.org

Preface

Air base defense and attack has been the subject of sustained research and analysis at the RAND Corporation for most of the history of the organization; RAND produced 264 reports on the topic between 1951 and November 2020. This report provides an overview of RAND's work over these seven decades and seeks to describe RAND's contributions (both in substance and analytical methods) and to identify enduring insights of interest to policymakers responsible for improving the resiliency of U.S. air bases in the face of modern threats.

This report is a product of the RAND Project AIR FORCE (PAF) continuing program of self-initiated research. Support for this research was provided by the research and development provisions of PAF's contract with the U.S. Air Force. The study described in this report was administered at the unit level within PAF. This research should be of interest to officials in the U.S. Air Force, other services, combatant commands, and the U.S. Department of Defense, as well as to those in the broader defense policy community.

RAND Project AIR FORCE

RAND Project AIR FORCE (PAF), a division of the RAND Corporation, is the Department of the Air Force's (DAF's) federally funded research and development center for studies and analyses, supporting both the United States Air Force and the United States Space Force. PAF provides DAF with independent analyses of policy alternatives affecting the development, employment, combat readiness, and support of current and future air, space, and cyber forces. Research is conducted in four programs: Strategy and Doctrine; Force Modernization and Employment; Manpower, Personnel, and Training; and Resource Management. The research reported here was prepared under contract FA7014-16-D-1000.

Additional information about PAF is available on our website:
www.rand.org/paf/

This report documents work originally shared with DAF in September 2020. The draft report, issued in September 2020, was reviewed by formal peer reviewers and DAF subject-matter experts.

Contents

Figures

Tables

Summary

Issue

From the dawn of the air power age to today, airfields have been recognized as essential military facilities, housing aircraft and the infrastructure needed to conduct air operations. Given this, combatants in major and minor wars have gone to great lengths to destroy enemy aircraft on the ground (where they are most vulnerable) and to deny the use of airfields through attacks on runways, fuel storage, and other supporting assets. At the same time, combatants have sought to protect their own air forces through active and passive defenses. U.S. Department of Defense (DoD) and U.S. Air Force (USAF) attention to air base defense and attack (ABD/A) has ebbed and flowed over time with changes in the broader strategic environment. As of 2020, the vulnerability of U.S. forward air bases is once again among the threats of greatest concern to USAF and DoD leaders. The RAND Corporation has worked on issues related to analyzing ABD/A for 70 years. Analysts and leaders actively working on air base resiliency will be most familiar with the results of current and ongoing RAND analyses—which are typically closely held. This report documents and highlights RAND's many contributions to the analysis of ABD/A over time and identifies enduring lessons that go beyond the particulars of time, place, and technology.

Approach

Our research approach leaned heavily on RAND's library and archival resources. We used RAND's online databases and print indexes to compile an initial list of reports related to ABD/A; after collecting a critical mass, we were able to find even more reports by mining the bibliographies of the initial set. This method of snowball sampling leaves us fairly confident that our bibliography of 264 reports is very nearly comprehensive. This set of 264 reports includes both classified and unclassified work. For the analysis of RAND publications, we included all 264 reports, but for the broader discussion of RAND work over the decades, we focused only on the subset of unclassified reports.

Conclusions

Analysis of ABD/A Contributions over 70 Years of Analysis

- Over seven decades, RAND analysis has responded to an ever-evolving geopolitical, military, and technological landscape in step with its DoD and USAF sponsors (Table S.1), but, given RAND's charter, it was not entirely bound by them.

- At critical junctures, RAND led its DoD and USAF sponsors, identifying emerging threats to air bases and potential solutions well before the broader community acknowledged them.
- RAND's greatest contributions were in its disciplined and creative application of more formal analytical tools to the problem of ABD/A. RAND researchers invented and applied these tools so that the relative utility of various offensive and defensive concepts could be measured systematically.

Broader Lessons Learned About Air Base Defense and Attack from RAND Analysis

- **Air bases have always been, and are likely to remain, priority targets in wars.** This is true for two reasons. First, modern air power has proven to be a versatile and essential element of military power, one that, at minimum, must be countered to prevail in conflict. Second, unlike navies and armies, which generate combat power from the fleet at sea and maneuver forces in the field, air forces generate combat power from fixed airfields.
- **Air base attackers will rarely limit themselves to a single attack mode.** In conventional wars, combatants have attacked airfields with aircraft, cruise missiles, naval gunfire, artillery, mortars, rockets, commando raids, armored forces, and drones. Most combatants—even insurgents—have multiple options for attacking airfields and will use them as conditions dictate.
- **There are no simple or cheap means to defend air bases.** A review of RAND research findings and the longer history of ABD/A offers no panacea to the problem of airfield vulnerability. No broad category (passive or active defense) offers perfect protection, nor is either category consistently the most cost-effective option. Similarly, no single type of active defense (e.g., fighter interceptors versus ground-based surface-to-air missiles[SAMs]) or passive defense (e.g., hardened shelters versus dispersal) offers complete protection or is reliably the most cost-effective.
- **Aircraft dispersal on and across bases has renewed salience for air base defense.** Aircraft dispersal arguably achieved maximum salience in the 1950s, when the vulnerability of the U.S. bomber force to nuclear attack became one of the nation's most urgent defense problems. Although not cost-free, dispersal was an option that could be implemented relatively quickly, certainly compared with building nuclear-hardened shelters or deploying active defenses at every base. As the standoff missile threat increased in the 2010–2020 period, various concepts for distributed operations have again regained prominence as among the most versatile and executable. Distributed operations present a host of challenges for the USAF but, on the whole, are often easier to implement than other passive defense options and do not require massive investments in infrastructure at bases that may not be needed in the next war.
- **ABD/A is best understood from a systems perspective.** Under conditions of uncertainty, planners must assess the performance of a range of defensive options against an even wider range of enemy offensive options. Systems analysis can help planners understand how key air base processes work. An air base takes inputs (e.g., aircraft, personnel, fuel, munitions) and then follows formal procedures and protocols (e.g., mission planning, aircraft maintenance, fueling, arming) to create usable products (aircraft manned and ready for missions), which can be measured using output metrics (e.g., sorties generated, enemy aircraft shot down per mission, targets struck).

Looking to the Future

"The Battle of the Airfields" will likely look quite different in the coming decades, but if history is any guide, RAND will continue to be actively involved in supporting USAF and DoD efforts to ensure the resiliency of American air power—whether that air power comprises mobile missiles, unmanned aircraft launched from trucks, or manned aircraft flying from more traditional air bases.

Table S.1. Highlighted RAND Contributions to Air Base Defense and Attack over 70 Years

Time Period	Highlighted RAND Contributions to ABD/A
1950–1959 Nuclear Threats to USAF Bases in the United States and Europe	• Foundations of deterrence theory • Systems analysis of bomber basing • Comprehensive analysis of vulnerability of tactical air bases in NATO • Analysis of potential contributions of SAMs to air base air defense • Analysis of aircraft and missile vulnerability during flyout • Cost-effectiveness analysis of hardened aircraft shelters
1960–1969 A Shift Toward Conventional and Offensive Operations	• Engineering analysis of hardened aircraft shelter designs • Analysis of runway attack tactics and weapons choices • Viability of conventionally armed IRBMs in air base attack • Application of Vietnam lessons learned to air base ground defense in Thailand
1970–1989 Conventional Warfare in Central Europe	• Explored complex dynamics and trade-offs of ABD/A scenarios using nascent modeling and simulation techniques • Assessed novel concepts for the use of remotely piloted vehicles • Recommended dispersing resources from main operating bases and designing future aircraft to be forward-deployable
1990–2009 Era of Rear Area Sanctuary for USAF	• Analysis of air base ground attack as an adversary asymmetric strategy • Comprehensive history of ground attacks on air bases • Detailed analysis of GPS-guided missile threat to USAF bases • Integration of missile attacks on air bases in a campaign-level model
2010–2020 Anti-Access Threat to U.S. Bases Reinvigorates Analysis of Air Base Defense	• Major advances in analytical methods for assessing conduct of operations in contested, degraded, and operationally limited environments • Comprehensive open-source assessment of the U.S.-China military balance, including relative ABD/A capabilities • Lessons learned from air base attacks during 26 conflicts • Assessment of force presentation implications of distribution air operations • Analysis of implications of adaptive basing concepts for Agile Combat Support

NOTE: Contribution can be from a single report or multiple reports.

Acknowledgments

Thanks to Obaid Younossi, Robert Tripp, and Jacob Heim for supporting a preliminary investigation of insights from RAND's earlier work on ABD under the auspices of the fiscal year (FY) 2018 Combat Operations in Denied Environments (CODE) Lessons Learned effort. Ted Harshberger funded the research documented in this report, conducted primarily in FY 2019. We thank him for the support and for his helpful guidance and feedback along the way. Thanks to Stacie Pettyjohn and Marc Robbins for helpful comments on earlier drafts. Jacob Heim helped inform our discussion of analytical methods. Thanks to reviewers Chris Bowie and Michael Lostumbo for their constructive criticism and practical suggestions for improving the report.

Walter Nelson, RAND Library manager, helped us identify and gain access to the large number of RAND reports on this topic. His work over many years with DoD archivists to declassify old reports should be noted as well, since it provided the foundation for this open source report. He also worked with DoD archivists to declassify a half dozen reports that we were able to include in this report. We simply could not have done this analysis without his deep knowledge of RAND's publication history. We greatly appreciate his generous assistance and enthusiasm for the effort. In addition to the recently declassified reports discussed publicly for the first time in this report, we also were able to get permission to use roughly 40 reports that had previously been restricted to RAND internal use only. Susan Scheiberg, associate director of the RAND Library, and Cara McCormick, manager of RAND's archives, played a vital role in determining which of these previously internal RAND documents could be publicly released. Teague Allen (supervisor, metadata and taxonomy) and Tristan Gable (metadata librarian) in RAND Knowledge Services helped find full names for RAND authors who (per convention during the earlier years) were shown with first initials only. RAND editor Phyllis Gilmore provided guidance on how to incorporate full names in citations.

Finally, thanks to Karin Suede for her help with the manuscript and to communication analyst Paul Steinberg, whose revisions (including Table S.1) greatly improved the report summary. Production editor Amanda Wilson supervised the publications process with her usual efficiency and good humor. Copyeditor James Torr polished our prose and ensured that the report met RAND's high standards for format and presentation. We thank them both for their important contributions to this effort.

We also want to acknowledge the dozens of unnamed RAND researchers and managers whose work helped shape and inform this field of study, from the first report in 1951 to those researchers involved in ongoing research and analysis. Without their sustained efforts over seven decades, we would have no story to tell. We regret that we were unable to discuss the details of research conducted by many of the RAND authors named in Appendixes A and B. We hope that this report will be updated in the future and that someday their many contributions to the field can be publicly acknowledged.

Abbreviations

A2/AD	anti-access/area denial
ABD/A	air base defense and attack
ABM	anti–ballistic missile
ACS	Agile Combat Support
C2	command and control
CC&D	camouflage, concealment, and deception
CEP	circular error probable
CODE	Combat Operations in Denied Environments
CONUS	continental United States
C-RAM	Counter–Rocket, Artillery, and Mortar
DoD	U.S. Department of Defense
FY	fiscal year
GPS	Global Positioning System
ICBM	intercontinental ballistic missile
IRBM	intermediate-range ballistic missile
ISR	intelligence, surveillance, and reconnaissance
JICM	Joint Integrated Contingency Model
MOB	main operating base
OSD	Office of the Secretary of Defense
PACAF	Pacific Air Forces
PAF	RAND Project AIR FORCE
PGMs	precision-guided munitions
RPV	remotely piloted vehicle
SAC	Strategic Air Command
SAM	surface-to-air missile
SLBM	submarine-launched ballistic missile
SOF	special operations forces
START	Strategic Tool for the Analysis of Required Transportation
TAB-ROM	Theater Air Base Resiliency Optimization Model
TAB-VAM	Theater Air Base Vulnerability Assessment Model
TAB VEE	Theater Air Base Vulnerability Evaluation Exercise
TSAR	Theater Simulation of Airbase Resources
USAF	U.S. Air Force

1. Introduction

Background

Airfields have long been recognized as military centers of gravity. From the earliest days of military air power, airmen have sought to attack enemy airfields while protecting their own. The first documented successful attack on an airfield occurred during the first months of World War I, when a Royal Navy Air Service aircraft destroyed a German Zeppelin at its base in Dusseldorf. Air bases have played a central role in warfare since World War II. The major combatants in World War II recognized the strategic importance of what historian Norm Franks termed the "Battle of the Airfields."[1] As a result, attacks on air bases figured prominently in major offensive operations during that war, most notably in the German 1940 offensive, the Japanese December 1941 air attacks on U.S. forces in Hawaii and the Philippines, and the German Operation Boddenplatte, a desperate 1945 attempt to regain initiative in the air war. Attacks on airfields have occurred in at least 25 other conflicts (involving both conventional and unconventional forces) since World War II and are likely to remain a priority target for years to come.[2] They have been attacked by aircraft, missiles, naval gunfire, artillery, mortars, rockets, commandos and, most recently, drones.[3]

Over the course of these many conflicts, airmen have gained countless insights on how best to protect aircraft on the ground, using active defenses (e.g., anti-aircraft artillery and surface-to-air missiles [SAMs]) and a host of passive measures (e.g., dispersal, camouflage, deception, hardening, and post-attack airfield recovery capabilities, such as runway repair).[4]

RAND's Early Involvement In Air Base Defense Analysis

When RAND was founded in 1946, the air base defense experience and lessons of World War II were fresh but did not appear immediately salient. RAND was focused on the potential contributions of advanced aerospace technologies to the nation's defense, exemplified by the

---

[1] Norman L. R. Franks, *Battle of the Airfields: Operation Bodenplatte, 1 January, 1945*, London: Grub Street, 1994.

[2] Insurgent and terrorist forces are also keenly aware of the benefits associated with attacking enemy air forces on the ground. A recent example is the January 5, 2020 Shahab attack on the U.S. airfield at Manda Bay, Kenya that killed three Americans and either damaged or destroyed six surveillance and medical evacuation aircraft. See Thomas Gibbons-Neff, Eric Schmitt, Charlie Savage, and Helene Cooper, "Chaos as Militants Overran Airfield, Killing 3 Americans in Kenya," *New York Times*, January 22, 2020.

[3] For more on the history of air base attacks and defenses see Alan J. Vick, *Air Base Attacks and Defensive Counters: Historical Lessons and Future Challenges*, Santa Monica, Calif.: RAND Corporation, RR-968-AF, 2015.

[4] For more on these defensive techniques see Vick, 2015, Chapter Five, and Alan Vick, Sean Zeigler, Julia Brackup, and John Speed Meyers, *Air Base Defense: Rethinking Army and Air Force Roles and Functions*, Santa Monica, Calif.: RAND Corporation, RR-4368-AF, 2020.

young institution's first publication (also in 1946): *Preliminary Design of an Experimental World-Circling Spaceship.*[5] But it wasn't so much the potential benefits of new aerospace technologies that led RAND to devote so much attention to the problem of air base defense, but rather the risks posed by future adversary acquisition of the most fearsome weapon yet devised—the atomic bomb. These risks were both at the strategic level (i.e., how to deter the use of nuclear weapons against the U.S. homeland) and the tactical/technical level (i.e., the vulnerability of key military targets to nuclear weapon effects).

The U.S. nuclear bombing of Hiroshima and Nagasaki in August of 1945 was immediately recognized by leading airmen as revolutionary, not just because of the war-ending offensive power of the atomic bomb but also because the threat of surprise air attack with nuclear weapons rendered obsolete previous defense strategies based on national mobilization over the course of months or years. Recognizing the profound changes that nuclear weapons would bring to the military enterprise, General Hap Arnold, Commander of the U.S. Army Air Forces, in October 1945 directed three senior airmen (Generals Carl Spaatz, Hoyt Vandenberg, and Lauris Norstand) to identify the implications of these weapons for the Air Force and nation. The Spaatz Board report, delivered to General Arnold a month later, stressed the requirement for a combat-ready air force on constant alert. Historian Phillip Meilinger summarized the report findings:

> The atomic bomb's awesome destructiveness meant than an enemy surprise attack could decide a war because there would be no time for mobilization. The United States must, therefore, maintain a strategic bombing force in being capable of either "smashing an enemy air offensive, or launching a formidable striking force." In short, the Air Force "on the alert" was to be America's new first line of defense—and offense—in the future.[6]

Writing in the February 1946 *National Geographic Magazine*, General Arnold expressed this new strategic reality in stark terms. In his article titled "Air Power for Peace," Arnold argued that the United States could no longer wait until a conflict to mobilize forces. Rather it would have to maintain a highly capable and ready force to deter war:

> It is our obligation, now and in the future, to organize our armed forces with the most modern weapons to secure the most powerful striking force at the least expense to the taxpayer. We must do this, not to prepare for another war, because such a catastrophe would almost certainly throw the whole world back for centuries if, indeed, it did not destroy our present civilization. We must do this to *prevent* another war—to perpetuate peace.[7]

That same year, Bernard Brodie, a Yale political scientist, edited the first book-length treatise on the atomic age. *The Absolute Weapon* laid the foundation for U.S. nuclear strategy and

---

[5] RAND Corporation, *Preliminary Design of an Experimental World-Circling Spaceship*, Santa Monica, Calif.: RAND Corporation, SM-11827, 1946.

[6] Phillip S. Meilinger, *Hoyt S. Vandenberg: The Life of a General*, Bloomington, Ind.: Indiana University Press, 1989, p. 63.

[7] H. H. Arnold, "Air Power for Peace," *National Geographic Magazine*, February 1946, p. 135.

deterrence theory. Brodie (who a few years later would conduct studies for the U.S. Air Force [USAF] while working at the RAND Corporation) famously captured the essence of the nuclear challenge, observing that "Thus far the chief purpose of our military establishment has been to win wars. From now on its chief purpose must be to avert them. It can have almost no other useful purpose."[8]

In the midst of these tectonic changes in military thinking and technologies, General Arnold suggested to Frank Collbohm of the Douglas Aircraft Company that a group of scientists and engineers be organized at Douglas to help the Air Force prepare for a future in which advanced technologies would play an increasingly large role in military strategy and operations. Collbohm agreed and on March 1, 1946, the U.S. Army Air Force (USAAF) signed a contract with Douglas Aircraft Company to "house an independent group of civilians" to assist the USAAF in future planning.[9] This effort was called Project RAND. Two years later, on May 14, 1948, the RAND Corporation was founded as an independent, nonprofit organization "dedicated to furthering and promoting scientific, educational, and charitable purposes for the public welfare and security of the United States."[10]

A little over a year later, on August 29, 1949, the Soviet Union shocked the world, detonating its first nuclear weapon.[11] Although the significance of future Soviet possession of nuclear weapons was recognized as early as 1946, few believed that the Soviet Union was sufficiently advanced in science and technology to develop its own nuclear weapons in the near term. The 1949 test consequently made defense against bombers armed with nuclear weapons a national priority. In particular, the need to ensure that the U.S. bomber force—the only means to deliver nuclear weapons at the time—could survive a surprise attack from the Soviet Union drove advances in early warning sensor networks, command and control (C2), and active and passive defenses. The primary purpose of these efforts was to ensure that no matter the size of an attack, sufficient numbers of bombers would have either launched under warning or survived to mount a devastating retaliatory attack on the Soviet homeland. The concept of assured retaliation quickly became foundational in American deterrence theory and defense policy.

RAND's creation and early years are inextricably tied to nuclear weapons and the beginning of the Cold War. RAND researchers played a unique and pivotal role in the development of new concepts, strategies, and policies related to nuclear weapons. As described by Fred Kaplan in *The Wizards of Armageddon*:

---

[8] Bernard Brodie, ed., *The Absolute Weapon: Atomic Power and World Order*, New York: Harcourt and Brace, 1946, p. 76.

[9] Fred Kaplan, *The Wizards of Armageddon*, New York: Simon and Schuster, 1983, p. 58.

[10] RAND Corporation, "A Brief History of RAND," webpage, undated.

[11] David Holloway, *Stalin and the Bomb: The Soviet Union and Atomic Energy, 1939–1956*, New Haven, Conn.: Yale University Press, 1994, p. 265.

> This is the RAND Corporation, and during the peak of the Cold War, most of its occupants did little but sit, think, talk, write, pass around memos, and dream up new ideas about nuclear war. Isolated from the hurly-burly of the rest of the world, the men and women (mostly men) of RAND nurtured an *esprit de corps*, a sense of mission, an air of self-confidence and self-importance. It was, in large measure, this atmosphere, this intoxication, that induced the gradual creation of a doctrine concerning nuclear weapons, nuclear deterrence, nuclear war-fighting; that identified this doctrine with RAND, and propagated the notion that "the RAND way" was the only legitimate way of thinking about the bomb.[12]

As RAND analysts wrestled with the twin challenges of nuclear warfighting and deterrence, they quickly recognized that the vulnerability of air bases to surprise nuclear attack was extremely dangerous and destabilizing. Under USAF sponsorship, RAND researchers began a large, sustained, and highly technical effort to (1) measure the blast, thermal, and radiation effects from nuclear strikes on parked aircraft, runways, and air base infrastructure and (2) identify the most effective means to mitigate or avoid this damage. In just nine years (from 1951 to 1959), RAND published more than 50 reports on the problem of air base (and strategic force) vulnerability to nuclear attack.

Although this report is focused on RAND research and analysis on the twin problems of air base defense and attack (ABD/A), it should be noted that the early RAND work on air base defense was motivated by the growing realization among defense policy elites that vulnerable nuclear forces invited attack and undermined deterrence. Because manned bombers were the only means to deliver nuclear weapons in the late 1940s and early 1950s, air base vulnerability to nuclear attack was a strategic problem of the highest order.

Several RAND staff members were instrumental in understanding this link and developing the nascent field of deterrence theory.[13] As noted above, Bernard Brodie was one of the first to articulate how nuclear weapons profoundly changed the international security environment. He would continue to make important contributions to deterrence theory and strategy during his RAND career. For example, Brodie's 1958 RAND report *The Anatomy of Deterrence* emphasized the importance of a survivable second strike force and explicitly discussed the relative merits of manned bombers vs. ballistic missiles.[14] Brodie's 1958 report was the foundation for the similarly titled Chapter Eight in his famous 1959 book *Strategy in the Missile*

---

[12] Kaplan, 1983, p. 51.

[13] For a comprehensive treatment of RAND's contributions to deterrence theory, see Austin Long, *Deterrence— From Cold War to Long War: Lessons from Six Decades of RAND Research*, Santa Monica, Calif.: RAND Corporation, MG-636-OSD/AF, 2008.

[14] Bernard Brodie, *The Anatomy of Deterrence*, Santa Monica, Calif.: RAND Corporation, RM-2218, 1958, p. 18.

Age, written while on staff at RAND and published by Princeton University Press.[15] RAND held the copyright and it appears to have been available at the time within RAND as R-335.[16]

Having spent the early 1950s leading a large multi-year RAND systems analysis of Strategic Air Command (SAC) basing options, Albert Wohlstetter was keenly aware of the many vulnerabilities of SAC's long-range bombers. His 1954 report *Selection and Use of Strategic Air Bases* offered almost 400 pages of detailed analysis of these vulnerabilities and of options to counter them. The report did not, however, explicitly discuss the strategic consequences of these vulnerabilities. A few years later, Wohlstetter did offer his thoughts on the strategic implications of vulnerable nuclear forces in his widely read and influential paper "The Delicate Balance of Terror" (first published as a RAND report in 1958, then as an article in *Foreign Affairs* in 1959).[17]

Economist Thomas Schelling was also developing foundational deterrence concepts. Schelling published *The Reciprocal Fear of Surprise Attack* as a RAND paper in 1958 and *The Threat That Leaves Something to Chance* as a RAND internal document in 1959.[18] The ideas in these reports became famous as Chapters Eight and Nine, respectively, in Schelling's Harvard University Press book *The Strategy of Conflict*.[19] Published in 1960, *The Strategy of Conflict* was a pathbreaking application of game theory to international relations that would go on to win the Nobel Prize in Economic Sciences in 2005.

Andrew Marshall was another member of this core group of analysts.[20] Marshall would go on to fame as the founder of the Office of Net Assessment in the Pentagon and its director for a remarkable 43 years. In 1959, he co-authored with Herbert Goldhamer a RAND report on deterrence of total war. One of the findings of this report was that a Soviet decision to attack the United States would be driven by their assessment of the vulnerability of SAC's bombers: "The likelihood that the Russians will choose total war is affected much more by their estimate of the

[15] Bernard Brodie, *Strategy in the Missile Age*, Princeton, N.J.: Princeton University Press, 1959.

[16] Continuing interest in the out-of-print Princeton edition led RAND in 2007 to make the original report available in PDF (available at no cost on the RAND website) and hardcopy (available for purchase on the RAND website and also via Amazon and other booksellers).

[17] See Albert Wohlstetter, *The Delicate Balance of Terror*, Santa Monica, Calif.: RAND Corporation, P-1472, 1958 and Albert Wohlstetter, "The Delicate Balance of Terror," *Foreign Affairs*, January 1959.

[18] Thomas C. Schelling, *The Reciprocal Fear of Surprise Attack*, Santa Monica, Calif.: RAND Corporation, P-1342, 1958, and Thomas C. Schelling, *The Threat That Leaves Something to Chance*, Santa Monica, Calif.: RAND Corporation, D(L)-6936, 1959.

[19] Thomas C. Schelling, *The Strategy of Conflict*, Cambridge, Mass.: Harvard University Press, 1960.

[20] For more on Marshall's life, see Andrew F. Krepinevich and Barry D. Watts, *The Last Warrior: Andrew Marshall and the Shaping of Modern American Defense Strategy*, New York: Basic Books, 2015; Sharon Weinberger, "The Return of the Pentagon's Yoda," *Foreign Policy*, September 12, 2018, and Julian E. Barnes, "Andrew Marshall, Pentagon's Threat Expert, Dies at 97," *New York Times*, March 26, 2019.

proportion of SAC that will survive their initial attack than by their estimates of what target system SAC will use in its retaliatory strike."[21]

Finally, Herman Kahn, published *Some Specific Suggestions for Achieving Early Non-Military Defense Capabilities and Initiating Long-Range Programs* as a RAND research memorandum in 1958.[22] Unlike the other RAND analysts, who were focused on ensuring the survivability of the U.S. nuclear retaliatory force, Kahn did not believe this was sufficient to deter a nuclear attack by the Soviet Union. Kahn argued that a wide range of civil defense measures were needed to ensure that the United States could survive and recover after nuclear strikes. In short, he envisioned invulnerable nuclear forces *and* an exceptionally hardened civil society and economy that could bounce back from massive destruction and the deaths of millions.[23]

For example, Kahn argued that key elements of American industry should be located in mines to ensure their survival after a nuclear war. These civil defense ideas were most fully developed in his 1960 Princeton University book *On Thermonuclear War*,[24] which brought him fame but also notoriety for his ideas and the blunt, antiseptic, and often flip way that he presented them. His fame was such that the *London Times* described Kahn as "the prototype for Dr. Strangelove."[25] Indeed, filmmaker Stanley Kubrick got the idea for the Doomsday Machine and "mineshaft gap" depicted in *Dr. Strangelove* straight from Kahn's book.[26] Kahn left RAND in 1961 to found the Hudson Institute. According to one biographer, Kahn left because of disputes with RAND leadership over his ideas on civil defense. His departure was certainly enabled by the fame that *On Thermonuclear War* brought him.[27]

Although the scale of effort devoted to air base defense in the 1950s would not be matched in following decades, ebbing and flowing with changes in the strategic environment, RAND analysts have continued to wrestle with the tactical and strategic aspects of ABD/A, publishing 264 reports between 1951 and November 2020. Indeed, since 2010 there has been a renaissance in this field due to Chinese fielding of precision long-range strike weapons (primarily ballistic

---

[21] Herbert Goldhamer and Andrew W. Marshall, with the assistance of Nathan Leites, *The Deterrence and Strategy of Total War, 1959–1961: A Method of Analysis*, Santa Monica, Calif.: RAND Corporation, RM-2302, 1959, p. 6.

[22] Herman Kahn, *Some Specific Suggestions for Achieving Early Non-Military Defense Capabilities and Initiating Long-Range Programs*, Santa Monica, Calif.: RAND Corporation, RM-2206-RC, 1958.

[23] For more on Kahn's thinking regarding invulnerable nuclear forces, see Sharon Ghamari-Tabrizi, *The Worlds of Herman Kahn: The Intuitive Science of Thermonuclear War*, Cambridge, Mass.: Harvard University Press, 2005, p. 215.

[24] Herman Kahn, *On Thermonuclear War*, Princeton, N.J.: Princeton University Press, 1960.

[25] *Sunday Times* (UK), February 9, 1965, cited in Ghamari-Tabrizi, 2005, p. 41.

[26] For discussion of links between Kubrick's film *Dr. Strangelove* and Kahn's work, see Ghamari-Tabrizi, 2005, pp. 274–280.

[27] Ghamari-Tabrizi, 2005, pp. 308–309.

missiles). Between 2010 and November 2020, RAND published 48 ABD/A reports, the most since the 1980s.

Purpose of This Report

This report seeks to document RAND's many contributions (both substantive and in advances in analytical methods) and to identify major insights and lessons from this body of work. The focus is on those enduring lessons that go beyond the particulars of time, place, and technology. USAF and DoD analysts and leaders actively working on air base resiliency will be most familiar with the results of current and ongoing analyses—which are typically closely held. This report is intended to supplement those narrower and more sensitive findings with a broader look at the more generalizable findings from decades of analysis.

Research Approach

Our research approach leaned heavily on RAND's library and archival resources. The online databases and print indexes at our disposal helped us to compile an initial list of reports related to ABD/A, and, after collecting a critical mass, we were able to find even more reports by mining the bibliographies of the initial set. This method of snowball sampling leaves us fairly confident that our bibliography of 264 reports is very nearly comprehensive.

To categorize the reports, we created a spreadsheet template to qualitatively code reports based on, for example, which threats and regions the report discussed. We also included basic report characteristics, such as the author, year of publication, a short description of the report's content, and the policy problem it addressed. These sheets were often completed by hand and then later compiled into a digital spreadsheet. The coding was often reviewed by more than one person for consistency and accuracy. After coding, we were able to quickly count the number of reports that existed in each category and decade, and then represent them in the figures shown in Chapter 2. Note that some reports span multiple topics within a broader category, so the counts will not add up exactly to the total number of reports.

In addition to generating descriptive statistics on the content of reports, we used these data to analyze the frequency of different RAND authors in our bibliography and the connections that exist among co-authors. The networks are represented graphically in Appendix A as "circos" plots and network graphs. Both kinds of visualizations were created using the Python programming packages networkx and nxviz.

Although the statistical and network analyses represent all of the reports in our dataset, the content in Chapters 3–6 pertains only to reports that were either unclassified at the date of publication or have since been declassified and approved for public use. Of the 264 total reports, 95 are unclassified or declassified. Forty of these latter reports had been either classified or "RAND internal use only" prior to our project. With the help of RAND librarians, we were able to get these reports cleared for public release. We believe that these are representative of the larger population but note some limitations. For example, almost all the RAND analysis conducted since 2010 is sensitive and not available to the public. Thus, although the newer findings are broadly consistent with insights from previous decades, the specifics cannot be discussed in this report.

Organization of This Report

In Chapter 2, we present an overview and descriptive statistics of RAND work published between 1951 and November 2020. In Chapter 3, we discuss RAND work published in the 1950s. This period was dominated by the nuclear threat to SAC bases in the continental United States. In Chapter 4, we discuss RAND's work in the 1960s, which initially retained the focus on SAC vulnerability to nuclear attack but in 1963 shifted away from nuclear threats to focus on conventional attacks on forward bases. In Chapter 5, we discuss reports published during the 1970s and 1980s, when the focus was on conventional warfare in Europe. In Chapter 6, we consider the 1990s and 2000s, a period when the United States enjoyed a period of conventional military dominance with few concerns about adversary attacks on U.S. bases. The 2000s were also marked by 9/11 and major counterinsurgency operations in Afghanistan and Iraq. In Chapter 7, we discuss RAND work published between 2010 and 2020, a period of renewed focus on air base vulnerability. Finally, in Chapter 8 we present research findings and enduring themes from RAND research published over this 70-year period.

2. A Statistical Overview of RAND Research on Air Base Attack and Defense, 1951–2020

The body of RAND research on attacking and defending air bases spans nearly 70 years and includes contributions from at least 260 different authors across more than 260 classified and unclassified reports. To better characterize such a large collection of work, in this chapter we present a set of statistical findings that reveal the most common research themes on ABD/A at different points in the organization's history. These statistical findings are drawn from the entire body of RAND work—both classified and unclassified. Later chapters will build on these insights by exploring specific eras in greater detail.[28]

Publication Rate and Series Type

With the help of RAND's archival resources and library support staff, we identified 264 total reports dating back to 1951 dedicated to ABD/A.[29] Of these reports, more than one-third were either unclassified at the time of original publication or have since been declassified.

While the frequency of publications on the subject has varied over time, no single time period dominates the others. Figure 2.1 shows that the publication rate remained fairly steady from 1951 to 1991. In those forty years, there were no fewer than ten reports released in any given five-year period.[30]

However, with the fall of the Soviet Union, the volume of reports significantly declined. Influential research was still published in this period, but the general consensus in the USAF community was that adversaries did not have the missile technology, air power, or special operations forces necessary to pose a credible threat to U.S. air bases. Not until the 2010s, which have been strategically defined by the anti-access/area-denial (A2/AD) threat posed by China and Russia, did RAND's publication rate on the subject return to its former Cold War levels.

---

[28] Subsequent historical narrative chapters only discussed unclassified (including declassified) reports.

[29] The bibliography is current up to November 2020. For our purposes, reports published in 2020 count toward the 2010s decade.

[30] RAND published just ten reports on ABD/A from 1977 to 1981.

Figure 2.1. RAND Reports on Air Base Defense/Attack, by Decade

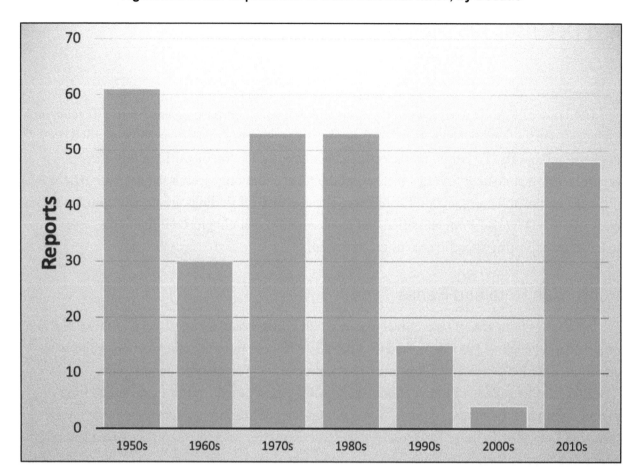

Although audiences outside of RAND staff are not likely to be familiar with its product slate and the naming conventions used for publications, the many different types of reports in our dataset are worth discussing briefly to properly understand later findings. At RAND, various designations are used to distinguish between products that have different dissemination restrictions, primary audiences, and technical review requirements, and these have evolved over time.[31] Our database includes all types of RAND reports, including working notes and other products intended for internal use only, draft reports that were delivered to clients but for various reasons did not become final reports (e.g., a draft was incorporated into another report), and peer-reviewed final products.

In our dataset, 120 (45 percent) of the reports were peer-reviewed and professionally edited final reports; 144 (55 percent) were internal or draft reports. Although the reports in our collection did not undergo precisely the same publication process, all report types will be counted equally in the statistical findings that follow.

---

[31] Over the seven decades covered by this report, RAND used a variety of prefixes to designate peer-reviewed reports, including RM, N, R, RR, MR, and MG. Similarly, various internal and draft reports had prefixes of D, IN, PR, WN, WD, D(L), and DRR.

Central Themes

We partitioned the reports into decades. From the start, we suspected that there would be significant differences between the research themes in different time periods, but these assumptions were only validated after individually coding each report. The coding process involved summarizing reports by certain subtopics that would help to differentiate them at a high level. Specifically, we recorded information on the offensive and/or defensive nature of the analysis (Figure 2.2), the types of threats described (Figure 2.3), the geography (Figure 2.4), the active defense solutions (Figure 2.5), and the passive defense solutions (Figure 2.6).

Our first finding was that the vast majority of reports (78 percent) were about defending U.S. and ally air bases.[32] Aside from the 1970s, every decade had more reports on defending air bases than on attacking adversary bases. This is not terribly surprising, given that the USAF's combat power has and will likely continue to be concentrated at air bases. The USAF must understand how best to defend its own source of combat power before considering how to attack adversary air bases, which are just one of many potential targets.

Figure 2.2. Number of Reports Focused on Attacking Adversary Air Bases, Defending U.S. Air Bases, or Both

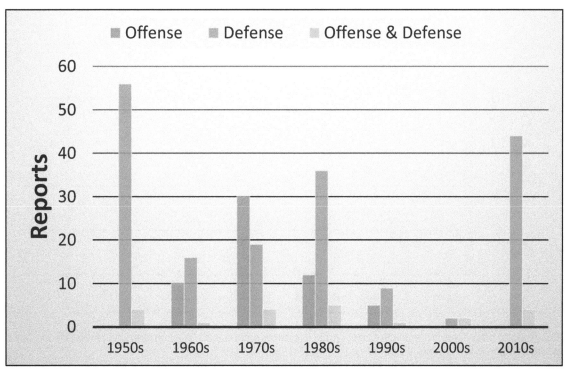

Defense was a mainstay of RAND Research on ABD/A from 1951 to 2020, but the type of threat and the regional focus varied often. Perhaps the most dramatic shift occurred between the

---

[32] This percentage includes reports about just defense as well as reports about both offense and defense.

1950s and 1960s. In the 1950s, most reports (32 in total) were devoted to protecting air bases from nuclear attacks in the continental United States (CONUS). As discussed at length in Chapter 3, the survival of the strategic bomber force at SAC bases in CONUS was a great concern during the 1950s. The bomber force was the sole nuclear deterrent at the time, and there was a fear that a surprise Soviet attack on those bases could dismantle the U.S. second-strike ability. Policymakers were especially interested in the vulnerability of these bases, since there was not yet a robust network of early warning radars and airborne interceptors to protect them. The research focus remained on the nuclear threat in CONUS until about 1961, but as the tenets of nuclear deterrence began to take shape, reports in the 1960s increasingly addressed the conventional threat instead. The Kennedy administration's Flexible Response strategy acknowledged that as the Soviet conventional threat grew stronger, massive retaliation as a deterrent to conventional attack became less credible, and additional nuclear and conventional response options were required (see Chapter 4).

Figure 2.3. Number of Reports on Different Types of Threats to Air Bases

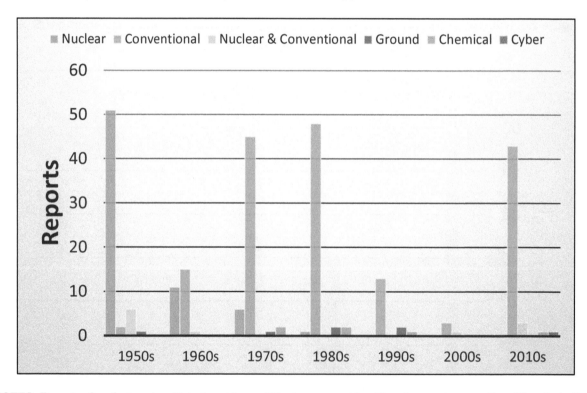

NOTES: Reports often featured multiple threat types. We show a combined "nuclear and conventional" bar, but other mixed threats appear in separate bars.

Reports in the 1970s were largely focused on conventional threats in Europe, but what made the decade particularly distinct was its emphasis on offensive counterair operations. Once again, this was most likely a result of the growing conventional threat presented by the Warsaw Pact. RAND research on attacking adversary bases sought to examine ways to disable Warsaw Pact airpower and blunt its ability to successfully execute a blitzkrieg-style attack across the continent (see Chapter 5). A total of 21 reports in the 1970s were specifically on offensive conventional operations in Europe. Reports in the 1980s still were about conventional threats in Europe, but they turned back to defensive rather than offensive operations.

In the decades that followed, threats outside the nuclear and conventional categories—such as ground, chemical, and cyber—appeared in small numbers, but the conventional threat still dominated. The same can be said about regions outside Europe and CONUS. The one exception to this is Asia, which has been the most frequently discussed region in the 2010s, with China being the primary adversary. The rapid advance of the People's Liberation Army's (PLA's) conventional missile capabilities in quantity, range, and precision has introduced new challenges to air base defense and U.S. force posture more generally in the Pacific (see Chapter 7).

Figure 2.4. Number of Reports by Regional Focus

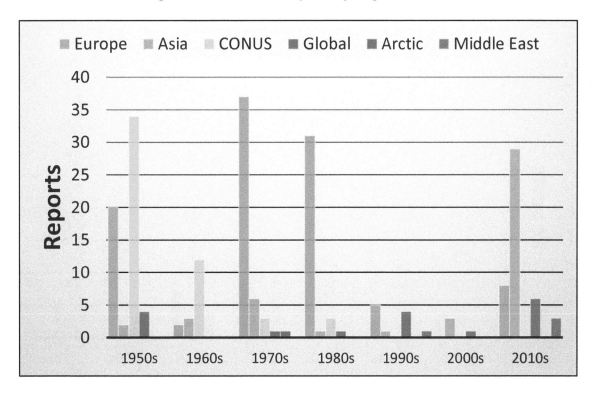

The next topic we tracked was active point defense, which we coded as either ground defense, ballistic missile defense, air defense, or counter–rocket, artillery, mortar (C-RAM). Figure 2.5 shows that active defenses have received more research attention in recent years than in the past, but in general they appear in our dataset far less often than passive defenses. This can partly be explained by our choice not to include reports on theater-wide air and missile defense, which include early warning radars and C2 nodes. Although the protection of critical assets at air bases is often central to the mission of theater air and missile defense assets, we did not include these kinds of reports in our bibliography unless they examined point defense at the base level.[33] It should also be noted that ground-based air and missile defense is an Army mission, which may have contributed to the lower number of reports on active defense.

Research on passive defenses has been numerous and varied. We coded nine total passive defense strategies, but the top four most occurring strategies have consistently been sheltering aircraft (hardening), dispersal of aircraft across bases, dispersal of aircraft on a single base, and recovery after attack. The distribution of these passive defense measures has remained remarkably consistent over the decades. In total, hardening appears most often, with 92 instances, but dispersal across bases and recovery are not far behind, at 72 and 75, respectively. While the cases for investing in hardening, dispersal, and recovery may have changed over time, no single strategy has dominated or proven superior.

Figure 2.5. Number of Reports That Investigated Active Point Defense Solutions

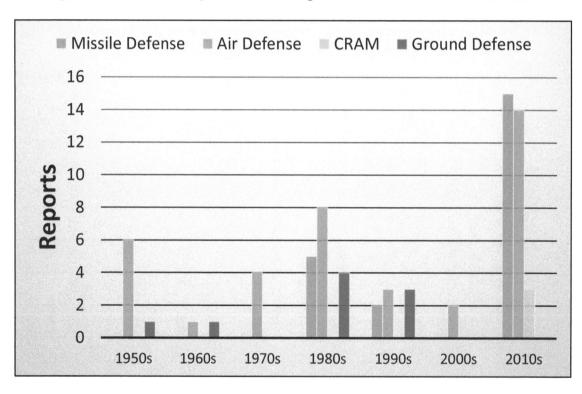

Other passive defenses that we coded in smaller numbers were camouflage, concealment, and deception (CC&D), launching aircraft from bases to protect them after early warning of an

---

[33] Here we make a distinction between RAND studies on active point defense solutions—such as anti-aircraft guns and short-range surface-to-air missiles—and studies related to longer-range systems that are not deployed exclusively to defend an air base. Modern examples of active point defenses are the Iron Dome and Indirect Fire Protection Capability systems. Modern examples of area and theater air defense systems are the Patriot and Terminal High Altitude Area Defense systems. The latter two, which rely on complex networks of sensors and shooters integrated across a region, can protect a variety of critical assets and are rightly treated as a separate research topic.

attack, building more runways at bases, and airfield battle damage repair. Additionally, when reports investigated novel ideas to provide logistics and maintenance support, we coded them as "new support concepts."

Figure 2.6. Number of Reports That Investigated Passive Defense Solutions

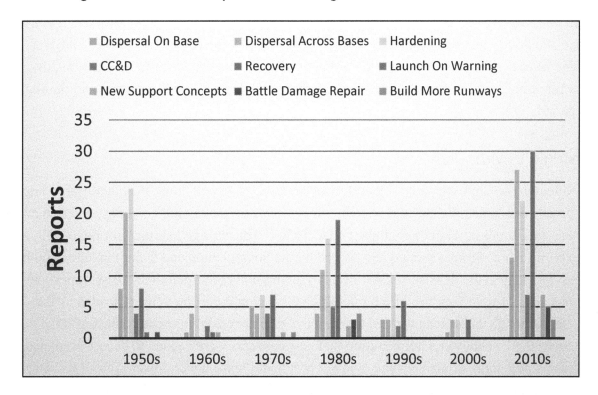

In the following chapters, we will discuss many of these themes in greater detail and cite influential reports that helped flesh out the concepts and propel further research in their direction.

3. Nuclear Threats to U.S. Air Force Bases in the United States and Europe, 1950–1959

In this chapter, we discuss RAND's first contributions to the study of air base defense and attack. We briefly describe the strategic environment during the early years of the Cold War, provide an overview of RAND research and analysis published between 1950 and 1959, discuss a few representative reports, and then conclude by highlighting the major themes and findings of this period.

Strategic Environment

Although the Cold War was well underway between 1946 and 1949, President Harry Truman was greatly concerned that defense spending was a threat to the overall economy, and as a result postwar cuts to military forces continued up to 1950.[34] This changed abruptly in June 1950, when North Korean forces (armed and trained by the Soviet Union and China) invaded South Korea. To many in the defense community, this aggression confirmed their worst fears about the ultimate aims of what they saw as a monolithic global communist movement headed by the Soviets. Although the Korean war led to a reversal of planned defense cuts, that conflict convinced the Eisenhower administration that new concepts were needed to deter communist aggression both more effectively and at lower cost in dollars and manpower. Thus, President Dwight Eisenhower and Secretary of State John Foster Dulles advanced the concept of massive retaliation to deter nuclear attack and, more ambitiously, avoid all wars through the threat of extensive nuclear strikes against the adversary. This threat was closely linked to the administration's New Look strategy that sought to make national defense affordable through a reliance on nuclear forces (at that point exclusively in the USAF) rather than large and more costly conventional ground forces.[35]

The USAF's Strategic Air Command was arguably the most important and powerful military organization during these years, growing at a remarkable rate.[36] For example, SAC went from 9

---

[34] For more on the beginnings of the Cold War during the Truman Administration, see Melvyn P. Leffler, *A Preponderance of Power: National Security, the Truman Administration, and the Cold War*, Palo Alto, Calif.: Stanford University Press, 1992.

[35] Robert R. Bowie and Richard H. Immerman, *Waging Peace: How Eisenhower Shaped an Enduring Cold War Strategy*, Oxford, UK: Oxford University Press, 1998, pp. 199–201; Richard M. Leighton, *History of the Office of the Secretary of Defense, Volume III: Strategy, Money, and the New Look, 1953–1956*, Washington, D.C.: Historical Office, Office of the Secretary of Defense, 2001, pp. 205–227.

[36] For more on SAC's early years, see Phillip S. Meilinger, *Bomber: The Formation and Early Years of Strategic Air Command*, Maxwell Air Force Base, Alabama: Air University Press, 2012; Walton S. Moody, Jacob Neufeld, and R. Cargill Hall, "The Emergence of the Strategic Air Command," in Bernard Nalty, *Winged Shield, Winged*

understrength bomber groups in 1946 to 38 full-strength bomber wings in 1955 and from 18 bases to 51 bases in the same time period.[37] As discussed in the introduction to this report, ensuring that a sufficient number of SAC bombers would survive a surprise attack and be able to launch a retaliatory strike was viewed as the foundation of a robust deterrent and, therefore, the highest strategic priority. Although it would be years before the Soviet Union possessed enough nuclear weapons and long-range aircraft able to deliver them, American defense leaders and planners recognized that it would also take years for the United States to devise, program, fund, and deploy the network of radars, communication centers, and several thousand fighter interceptors necessary to detect and intercept Soviet bombers. There was a great sense of urgency, driven by a sense that the military balance was shifting in favor of the Soviet Union, to the point that there was serious talk of a preventive war strategy.[38] In this policy and emotional environment, USAF efforts to reduce the vulnerability of SAC bombers received a level of senior leader attention far beyond that of any subsequent period.

The principal threat driving work on air base defense was the prospective threat of Soviet bombers delivering fission and fusion weapons in a surprise attack against U.S. bomber bases. Although the fundamental pillars of air base defense (active defense, hardening, dispersal, CC&D) were discovered during World War I, then refined and combat tested during World War II, the invention of atomic weapons meant that in a future war, American air bases could be subjected to blast, thermal, and radiation effects previously experienced only in the U.S. atomic bombings of the Japanese cities of Hiroshima and Nagasaki. The vulnerabilities of parked aircraft and air base infrastructure to such effects were not well understood at the beginning of the nuclear age. Planners did understand, however, that some bombers would get through U.S. air defenses and drop nuclear weapons on air bases. An understanding of these damage mechanisms was, therefore, necessary in order to fully grasp the breadth and depth of the nuclear threat to air bases. For these and other reasons, the United States embarked on an extensive nuclear weapon testing program of more than 1,000 detonations over almost 50 years. One hundred eighty-eight of these tests occurred between 1951 and 1958. All were conducted above ground, with many designed to understand effects on various targets.[39]

Sword: A History of the United States Air Force, Volume II, 1950–1997, Washington, D.C.: Air Force History and Museums Program, 1997, pp. 53–96; Henry Narducci, *Strategic Air Command and the Alert Program: A Brief History*, Offutt Air Force Base, Nebraska: Office of the Historian, Headquarters, Strategic Air Command, 1988; and Strategic Air Command, Office of the Historian, *Alert Operations and the Strategic Air Command: 1957–1991*, Offutt Air Force Base, Neb., 1991.

[37] Alan J. Vick, *Force Presentation in U.S. Air Force History and Airpower Narratives*, Santa Monica, Calif.: RAND Corporation, RR-2363-AF, 2018, pp. 25–27.

[38] For more on American government fears regarding Soviet nuclear capabilities, see Marc Trachtenberg, "A 'Wasting Asset': American Strategy and the Shifting Nuclear Balance, 1949–1954," *International Security*, Vol. 13, No. 3, Winter 1988–1989, pp. 5–49.

[39] U.S. Department of Energy, Nevada Operations Office, *United States Nuclear Tests: July 1945 Through September 1992*, Las Vegas, Nev., 2000, p. xi.

The 1950s tests provided a treasure trove of empirical data on nuclear effects. For example, aircraft and other military systems were placed at various distances from nuclear test detonations to assess blast, thermal, and other damage mechanisms.[40] RAND ABD/A studies of the 1950s (e.g., Sandoval's analyses of long-duration blast loading on shelter doors) were able to draw on these data to delve more deeply into the specific vulnerabilities of air bases.[41]

Policymaker interest in dispersal and hardening options remained high throughout the 1950s, triggering many technical studies, tests, and supporting analyses. For example, in 1952 the USAF's Joint Air Defense Board (JADB) began studying shelter options to protect aircraft from nuclear blast effects. In 1954, the Bomb Burst Committee recommended that on-base fuel, pump houses, communications facilities, radar transmitter buildings, and equipment shelters all be hardened to withstand blast effects from nuclear weapons. "The Air Force explicitly called out equipment shelters primarily as shelters for aircraft, with the intention of protecting large numbers of planes for retaliatory takeoff."[42] Early warning, interceptor aircraft, ground-alert for a portion of the bomber force, and dispersal (on and across bases) were ultimately chosen as the most cost-effective options for protecting bombers. Secretary of the Air Force Donald A. Quarles described these programs in his 1956 testimony before the Senate Armed Services Committee "Airpower" hearings:

> Adding to the growing capability of the United States Air Force to absorb an initial attack and strike back effectively is the dispersal program for Strategic Air Command bases in the United States. This program will be well under way in the coming fiscal year. And we will continue to improve the alert status of SAC bombers.[43]

Overview of RAND Research on Air Base Defense and Attack During This Period

As noted above, creating a North America air defense network was an essential and enormous undertaking. DoD embarked on a host of early warning and active defense measures during this period, including (in partnership with Canada) the creation of a line of early warning radars in Canada (the Distant Early Warning [DEW] Line, Mid-Canada Line, and Pinetree Line),

---

[40] Test results were first reported in the 1950 book *The Effects of Atomic Weapons*. Updated editions were published every few years as *The Effects of Nuclear Weapons*. See Samuel Glasstone and Philip J. Dolan, *The Effects of Nuclear Weapons*, Washington, D.C.: U.S. Departments of Defense and Energy, 1977. See pp. 194–195 and pp. 226–277 for effects against aircraft.

[41] Charles A. Sandoval, *A Handbook for Estimating Material Requirements and Costs of Shelter Doors Subjected to Long-Duration Blast Loading*, Santa Monica, Calif.: RAND Corporation, RM-2277, 1958.

[42] Karen Weitze, *Eglin Air Force Base, 1931–1991: Installation Buildup for Research, Test, Evaluation and Training*, Eglin Air Force Base, Fla.: Air Force Materiel Command, 2001, p. 231.

[43] U.S. Senate, "Study of Airpower," hearings before the Subcommittee of the Air Force of the Committee on Armed Services, Part XXI, 84th Congress, 2nd Session, June 26 and 28, 1956, p. 1542.

supplemented by naval and airborne early warning radars. Active defenses included the deployment of a large USAF interceptor force, development and deployment of Army Nike missiles around cities and SAC bases, and USAF deployment of Bomarc long-range surface-to-air missiles (SAMs).[44]

RAND was involved in active defense studies at this time, especially on early warning and C2—for example, design of the Semi-Automated Ground Environment (SAGE) system for air defense C2—but these efforts were investigating the problems of continental or, at minimum, area defense, not the point defense of air bases.[45] To be sure, the early warning and C2 improvements were of great consequence for SAC bomber survival, since they could greatly improve the effectiveness of airborne and ground-based active defenses, although these defenses were not expected to be sufficiently robust to prevent significant numbers of enemy bombers from leaking through and attacking bases. Rather, the strategic value of early warning and rapid C2 was to warn bomber bases of pending attacks. With adequate early warning, bombers could be kept on alert and launched prior to the arrival of enemy weapons—a concept that became central to U.S. deterrence theory and SAC operations during most of the Cold War. That said, in order to make the scope of this research and report manageable, we do not discuss these broader efforts in this report.

Additionally, most RAND ABD/A studies during this period did not attempt to include theater or continental air defense and warning within their analysis, focusing instead on passive defensive options to mitigate the effects of nuclear strikes, including dispersal of aircraft on and across bases and hardening of air bases.

As illustrated in Table 3.1, RAND ABD/A research during the 1950s reflected these national and Air Force priorities. Most RAND reports of this period sought to ensure the survival of the strategic nuclear force—at the time made up entirely of long-range bombers—by assessing the cost-effectiveness of various hardening and dispersal concepts. Many of these concepts (e.g., protecting bombers in underground shelters) were invented at RAND.[46] As Karen Weitze notes in her history of Eglin Air Force Base, RAND was deeply involved in USAF efforts to design nuclear-hardened shelters for SAC aircraft. "In 1957, Rand formally hired Weidlinger Associates

---

[44] For more on early Cold War efforts to defend the United States from air attack, see Kenneth Schaffel, *The Emerging Shield: The Air Force and the Evolution of Continental Air Defense 1945–1960*, Washington, D.C.: Office of Air Force History, 1991; Mark L. Morgan and Mark A. Berhow, *Rings of Supersonic Steel: Air Defenses of the United States Army 1950–1979: An Introductory History and Site Guide*, Bodega Bay, Calif.: Hole in the Head Press, 2010; and Mark Berhow, *U.S. Strategic and Defensive Missile Systems, 1950–2004*, Oxford, UK: Osprey Publishing, 2005.

[45] For more on RAND's role in SAGE, see Schaffel, 1991, pp. 155–161. An example of RAND research on active air defense is E. J. Barlow, *RAND Air Defense Analysis*, Santa Monica, Calif.: RAND Corporation, RM-562, 1951. (Declassified by the Air Force Declassification Office on August 14, 2015). Barlow led much of the RAND research on air defenses in the 1950s.

[46] See, for example, William M. Capron, *Let's Build a B-52 Shelter Now*, Santa Monica, Calif.: RAND Corporation, D-6159-PR, 1959.

to study the limits of blast-resistant steel aircraft shelters. Rand then hired Weidlinger in 1959 to design and engineer an alert shelter for SAC's small jet bomber, the B-58, stipulating that the shelter be resistant to nuclear blast effects."[47] These engineering studies directly supported RAND operational analyses but also ongoing USAF test and evaluation efforts to better harden structures.

To support these larger assessments of defensive options, RAND scientists published nine reports that quantified the damage mechanisms and effects of nuclear weapons on air bases, including parked aircraft, runways, hangars, fuel storage, and other infrastructure. Another seven reports explored means to conduct air operations during and immediately after a nuclear attack. These reports explored air base recovery challenges (e.g., debris removal from runways) and operational issues (e.g., flying through radioactive dust clouds) associated with the immediate aftermath of a nuclear attack. Five reports were devoted to some aspect of air defense against bombers, including early warning requirements, interceptor operations, and short-range air defense at bases. Three reports sought to understand the effects of nuclear strikes on the complex chain of events necessary to generate aircraft sorties, a more sophisticated measure that offered greater insight into the actual warfighting implications of attacks. In addition to the reports shown in Table 3.1, another six addressed various other aspects of air base defense, such as air defense and how to conduct operations during and after a conventional air attack.

Table 3.1. Top Policy/Analytical Objectives of RAND Reports on Air Base Defense and Attack: 1951–1959

Policy/Analytical Objective	Number of RAND Reports
Ensure strategic force survival	31
Quantify effects of nuclear weapons on air bases	9
Conduct air operations during and after nuclear attack	7
Defeat enemy air attacks	5
Quantify effects of nuclear attacks on NATO sortie generation	3

NOTE: The table only includes reports in the top five categories, not all reports in the decade.

A Sampling of Reports

The first RAND report to address air base vulnerability as a force planning and basing location problem was published in 1951. Authored by Albert Wohlstetter, *Economic and Strategic Considerations in Air Base Location: A Preliminary Review* described the basing problem from the perspective of total system costs, including vulnerability considerations and the costs of "such passive defense measures as dispersal of aircraft and preparation for repair of

---

[47] Weitze, 2001, p. 232.

air bases."[48] This effort continued over a multiyear period and resulted in two highly influential reports. In 1953, *Special Staff Report: The Selection of Strategic Air Bases*, a Top Secret, 32-page summary of the analysis, was delivered to the Deputy Chief of Staff, Development, Headquarters, USAF. In 1954, the full 430-page analysis was delivered to the USAF as *Selection and Use of Strategic Air Bases*.[49] This body of analysis became a model for comprehensive systems analysis, in which campaign objectives, adversary actions, force vulnerabilities, air base infrastructure, aircraft sortie rate, aircraft design (e.g., range, speed and payload), and acquisition costs were combined to identify the most cost-effective solutions. (We discuss RAND's contributions to systems analysis in greater detail later in this chapter.)

The Wohlstetter report considered three types of strategic basing. The first relied on overseas bases that were relatively close to their targets. The second type based aircraft in the United States and refueled them in the air. The third type based aircraft in the United States and refueled them on the ground at forward bases. The essential findings of the Wohlstetter report were that (1) the overseas basing system was overly vulnerable to surprise attack, (2) the homeland basing/air-refueling system was too expensive (and would consequently force the USAF into giving up bombers to buy tankers) and (3) the homeland basing/ground refuel abroad was the most cost-effective. The report summary stated that "Within the framework of this analysis, systems consisting of United States primary bases and overseas refueling bases appear markedly superior."[50]

USAF leaders were less concerned about the cost of the homeland basing/aerial refueling option, believing (rightly) that U.S. strategic nuclear capabilities were such a high priority for national leadership that the funding would be made available. Two other factors contributed to the USAF pursuit of tankers: (1) The Soviet detonation of its first H-bomb in 1954 greatly increased the vulnerability of the forward refueling bases proposed by RAND and (2) General Curtis LeMay's determination to deploy larger numbers of B-52 bombers and be freed from dependence on overseas bases.[51]

Because the USAF was currently using the first, highly vulnerable, system, the study proved enormously valuable in convincing USAF leaders and other key decisionmakers that the problem was urgent and required immediate action. The study also provided essential data regarding the operation of the air-refueling basing system. Although the RAND team initially concluded that this option was too expensive, the USAF decided to pursue air refueling as the most robust of the three.

---

[48] Albert Wohlstetter, *Economic and Strategic Considerations in Air Base Location: A Preliminary Review*, Santa Monica, Calif.: RAND Corporation, D-1114, 1951, p. 1.

[49] A J. Wohlstetter, F. S. Hoffman, R. J. Lutz, and H. S. Rowen, *Selection and Use of Strategic Air Bases*, Santa Monica, Calif.: RAND Corporation, R-266, 1954.

[50] Wohlstter et al., 1954, p. vi.

[51] Kaplan, 1983, p. 106.

Other RAND reports offered detailed cost-effectiveness analyses of the relative merits of dispersal on or across bases, hardening key infrastructure (particularly aircraft shelters), protecting fuel, and restoring runway use after nuclear detonations on or near the air base. A variety of aircraft shelter designs were analyzed to determine which could withstand blast effects from nuclear detonations. Figure 3.1 illustrates a RAND design for a B-58 alert shelter.

Figure 3.1. RAND B-58 Alert Shelter

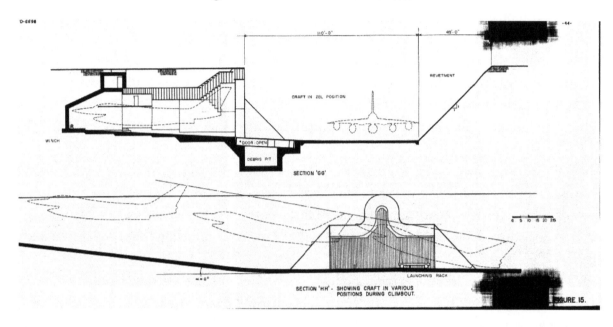

SOURCE: Paul Weidlinger, *An Alert Shelter for the B-58*, Santa Monica, Calif.: RAND Corporation, D-6698-PR, 1959, Figure 15.

Although the analytical emphasis during the early 1950s was on ensuring the survival of SAC bombers, there was a large and sustained effort to address the vulnerability of NATO tactical air bases to nuclear attack. The "Preservation of Tactical Air Command Potential in Western Europe" study produced 13 reports (all in 1954) on topics ranging from early warning to active and passive defense options. W. Baldwin and D. Davis's report *Wing-Level Defense Against A-Bombing* is representative of the analysis. This study identified measures that tactical air wings based in Europe could take to "reduce drastically the loss of combat potential as the result of enemy atomic attacks that may occur in a major war in the 1955 to 1958 period."[52] Operational effectiveness, cost, lead time, and strategic life of the proposed measures were all considered in the study.

---

[52] W. W. Baldwin and D. J. Davis, *Preservation of Tactical Air Combat Potential in Western Europe: Wing-Level Defense Against A-Bombing*, Santa Monica, Calif.: RAND Corporation, RM-1462, 1954, p. 1. (Declassified by Air Force Declassification Office on October 24, 2017.)

RAND Contributions

The fundamental concepts for air base defense date back to World War I, when active defenses (using general-purpose light and medium machine guns) and passive defenses (including dispersal, camouflage, and deception) were first introduced. During World War II, early warning networks, interceptor aircraft, and specialized anti-aircraft weapons increased the lethality of active defenses; dispersal, camouflage, and deception techniques were greatly improved, and hardening (especially aircraft revetments and shelters with overhead protection) of air bases made passive defenses much more effective.[53]

Thus, the conceptual foundation for air base defense was laid well before RAND researchers began to tackle the problem. There were, however, two new factors.

First, as noted above, nuclear weapon effects were so vast and different from prior weapons that the RAND researchers could not rely on historical experience to guide their assessments at either the strategic or tactical/technical level. As discussed in the introductory chapter, RAND analysts made major contributions to the nascent fields of nuclear strategy and deterrence theory that sought to inform defense and foreign policy in a nuclear-armed world. RAND analysts also made original contributions in understanding the tactical and operational implications of nuclear weapon use.

To understand the operational significance of nuclear weapons, RAND researchers used data from nuclear tests, where available, and made estimates where data were not available. The militarily-relevant nuclear effects went beyond the immediate blast, thermal, and radiation damage mechanisms to include a new temporal dimension. For example, one study explored a scenario in which sheltered missiles and aircraft had survived nuclear detonations at or near the base and were now launching retaliatory strikes. The researchers asked, "What is the vulnerability of these missiles and aircraft in flight to post-detonation nuclear effects?" The study identified threats to launching missiles and aircraft, including high wind velocities; dust, sand, stone, and debris density; and gamma radiation—all as a function of time and space.[54] This analysis helped determine how long after a detonation such systems would need to wait in their shelters in order to have a high probability of escaping undamaged. This study and related analyses would later help military planners understand the risks of "pin down" attacks, in which a less accurate but fast system (e.g., an early-generation intercontinental ballistic missile [ICBM]) would trap aircraft or missiles until slower but more accurate weapons (e.g., bombers) could arrive to destroy the aircraft in place.

---

[53] For an overview of these techniques and their use in past conflicts, see Vick, 2015.

[54] William M. Brown, *Vulnerability of Quick-Reacting Sheltered Missiles and Aircraft During Launch*, Santa Monica, Calif.: RAND Corporation, D-6625, 1959.

Second, RAND introduced new methods, such as cost and systems analysis, to assist military planners in understanding the relative merits of policy choices.[55] Cost analysis techniques allowed RAND researchers to explore how various classes of expenses varied across options. For example, a 1952 study on B-36 dispersal found that (no surprise) dispersing a wing across three bases was more expensive. What is interesting is that the cost driver was not the cost of additional runways and supporting facilities but the increased manning requirements.[56] By disaggregating costs into categories such as facilities, equipment, stocks, transportation, personnel, maintenance, and POL (petroleum, oil, and lubricants), analysts could determine cost drivers (which often were not intuitive) and, when combined with other factors, help determine the most cost-effective options, especially when cost analysis was used in support of systems analysis. Figure 3.2 illustrates the dispersal concept.

---

[55] For an overview of RAND contributions to management science, see Paul K. Davis, "Analytic Methods," in *Project AIR FORCE: 1946–1996*, Santa Monica, Calif.: RAND Corporation, 1996, pp. 47–51.

[56] RAND Corporation, Cost Analysis Section, *The Cost of Decreasing Vulnerability of Air Bases by Dispersal: Dispersing a B-36 Wing*, Santa Monica, Calif., R-235, 1952, p. 7.

Figure 3.2. Concept for Dispersing B-36 Bombers

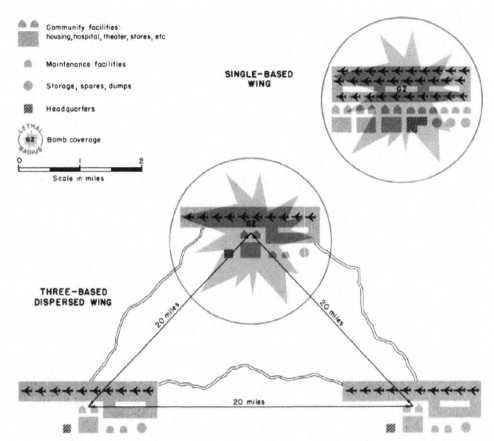

Fig. 1—Decreasing the vulnerability of an air base by dispersal

SOURCE: RAND Corporation, 1952, p. v.

According to Edward S. Quade, "Systems analysis, that is, analysis to suggest a course of action by systematically examining the objectives, costs, effectiveness, and risks of alternative policies or strategies—and designing additional ones if those examined are found wanting—represents an approach to, or way of looking at, complex problems of choice under uncertainty."[57] For air base defense studies, RAND researchers used systems analysis to consider the interplay of base location, vulnerability of infrastructure, effectiveness of defenses, effectiveness of attacking weapons, distance to targets, effectiveness of enemy air defenses, military objectives, defensive and offensive capabilities, and costs in assessing the relative utility of policy choices.[58] Advances in game theory, statistics and probability, data processing,

---

[57] E. S. Quade, *Military Systems Analysis*, Santa Monica, Calif.: RAND Corporation, RM-3452-PR, 1963.

[58] For early RAND thinking on analytical methods, see H. Igor Ansoff, W. W. Baldwin, D. J. Davis, Norman Maurice Kaplan, Paul Kecskemeti, and Albert Wohlstetter, *Outline of a Study for the Plans Analysis Section*, Santa Monica, Calif.: RAND Corporation, D-937, 1951; and Albert Wohlstetter, *Systems Analysis Versus Systems Design*, Santa Monica, Calif.: RAND Corporation, P-1530, 1958. These techniques continued to evolve at RAND as

simulation, and modeling created a virtuous circle in which an advance in one area spurred advances in others. RAND's decision in 1950 to design and build one of the most advanced computers in the country, the Princeton-class computer JOHNNIAC, was such an example, driven by the inability of RAND's existing computational devices to meet its analytical needs, as well as a shortage of commercially available systems with sufficient power. It is noteworthy that in 1950 RAND already had one of "the world's largest installations for scientific computing . . . [which] . . . operated six IBM 604 calculators around the clock" and had ordered two new IBM computers but, nevertheless, felt it needed to pursue even greater capabilities and even tackle the design and building of such a computer itself.[59]

Table 3.2 lists six of RAND's most prominent contributions to this field of study during the 1950s. The particular findings of these studies are too numerous to discuss here. We will, however, consider broader findings in the report's concluding chapter.

Table 3.2. Major RAND Contributions During the 1950s

RAND Contribution	Documented in
Foundations of deterrence theory	Brodie (1958), Schelling (1958, 1959), Wohlstetter (1958)
Systems analysis of bomber basing	Wohlstetter et al. (1954)
Comprehensive analysis of vulnerability of tactical air bases in NATO	Baldwin and Davis; Tuck; Skogstad and Snow; Stockton; all published in 1954
Analysis of potential contributions of SAMs to air base air defense	Tuck (1954)
Analysis of aircraft and missile vulnerability during flyout	Brown (1959)
Cost effectiveness analysis of hardened aircraft shelters	Stockton (1954)

In the next chapter, we will explore how the focus of RAND research broadened to include the analysis of conventional threats and, most significantly, began to consider how to attack enemy air bases to greatest effect.

---

described by Wohlstetter a decade later. See Albert Wohlstetter, *Theory and Opposed-Systems Design*, Santa Monica, Calif.: RAND Corporation, D-16001-1, 1968.

[59] F. J. Gruenberger, *The History of the JOHNNIAC*, Santa Monica, Calif.: RAND Corporation, RM-5654-PR, 1968, p. 2. A comprehensive treatment of RAND's contributions to computer science can be found in Willis H. Ware, *RAND and the Information Evolution: A History in Essays and Vignettes*, Santa Monica, Calif.: RAND Corporation, CP-537-RC, 2008.

4. A Shift Toward Conventional and Offensive Operations, 1960–1969

Strategic Environment

As the 1960s dawned, Soviet nuclear capabilities—both real and imagined—continued to dominate defense policy discussions. Among the more contentious debates was the claim advanced by some USAF generals and journalists (and supported to some degree by National Intelligence Estimates) of a "Missile Gap" in which the Soviet Union possessed a huge and growing lead in nuclear-armed ICBMs.[60] The controversy over Soviet nuclear force capabilities reflected the nascent (although rapidly improving) state of intelligence collection in the late 1950s and early 1960s, legitimate disputes over the evidence, bureaucratic politics among the services, and the lingering fear (exacerbated by the successful launch of Sputnik in 1957) that the Soviet missile design and production system was greatly superior.[61] It didn't help that a few months after Sputnik, a huge audience watched a U.S. missile test fail spectacularly, rising a few feet in the air, only to then descend and explode. Newspaper headlines ridiculed the test with terms such as "Flopnik," "Stayputnik," and "Kaputnik."[62]

In 1958 and 1959, syndicated columnist Joseph Alsop sensationalized the gap with a series of op-eds that appeared across the country in newspapers as diverse as the *Washington Post* and the *Eugene Register-Guard* in Eugene, Oregon. Alsop claimed that by 1960 the Soviet Union would have 100 ICBMs compared with 30 possessed by the United States and that by 1963 the Soviets would have over ten times as many missiles as the United States (1,500 versus 130).[63] Senator John F. Kennedy found Alsop's arguments compelling and, seeing the broad public interest in

---

[60] See Christopher A. Preble, *John F. Kennedy and the Missile Gap*, DeKalb, Illinois: Northern Illinois University Press, 2004, and Roy E. Licklider, "The Missile Gap Controversy," *Political Science Quarterly*, Vol. 85, No. 4, December 1970, pp. 600–615.

[61] RAND made major contributions in the area of technical intelligence collection which helped disprove the missile gap. RAND engineers designed a satellite reconnaissance system as early as 1951. In 1954, as part of Project FEEDBACK, RAND recommended that the USAF develop a satellite imaging system, and RAND's 1957 recommendations contributed to the ultimate design of the CORONA satellite reconnaissance system. See Michael D. Rich, *RAND's Role in the CORONA Program: Remarks on the 35th Anniversary of the First Successful Mission*, Santa Monica, Calif.: RAND Corporation, P-8017, 1998, pp. 1–2; Merton E. Davies and William R. Harris, *RAND's Role in the Evolution of Balloon and Satellite Observation Systems and Related U.S. Space Technology*, Santa Monica, Calif.: RAND Corporation, R-3692-RC, 1988, and Curtis Peebles, *High Frontier: The United States Air Force and the Military Space Program*, Washington, D.C.: Air Force History and Museums Program, 1997, pp. 5–7 and p. 12.

[62] Daniel Ellsberg, *The Doomsday Machine: Confessions of a Nuclear War Planner*, New York: Bloomsbury Publishing, 2017, p. 34.

[63] Joseph Alsop, "True Missile Gap Picture Belies Pentagon Response," *Eugene Register-Guard*, October 13, 1959.

the topic, began publicly criticizing the Eisenhower administration for its failure to address the gap, first in Kennedy's bid for reelection to the Senate in 1958 then again during his campaign for President in 1960.[64]

Both Alsop's claims and the less extreme National Intelligence Estimates later proved to be wildly off the mark. There was indeed a missile gap, but one that greatly favored the United States. In 1960, the United States possessed 12 ICBMS and the Soviets none. In 1961, the two countries' respective inventories totaled 63 and 4. By 1963, when Alsop had predicted a ten-to-one Soviet advantage, the actual numbers were 631 U.S. ICBMs to only 100 Soviet missiles.[65]

Although candidate Kennedy had been a proponent of the missile gap, once in office his administration quickly moved away from such claims. Of note for this report, Kennedy administration officials "who had come from the RAND Corporation, or who had close contacts with RAND, were apparently particularly skeptical" of the missile gap claims.[66] Only a month after Kennedy's inauguration in 1961, Secretary of Defense Robert McNamara publicly rejected the idea of a missile gap, but the USAF continued to maintain that the Soviet Union had as many as 800 ICBMs. During a Thanksgiving 1961 meeting with his defense advisors, Kennedy concluded that "the weight of evidence was clearly against the Air Force, and the issue finally withered away."[67]

The missile gap was debunked, but U.S. leaders and the public nevertheless feared Soviet nuclear potential, spurring efforts to rapidly deploy land-based ICBMs and sea-based submarine-launched ballistic missiles (SLBMs), to continue efforts to ensure the survivability of the strategic bomber force, and to improve the U.S. ability to collect intelligence on Soviet nuclear programs. The latter drove innovations in high-altitude airborne reconnaissance (e.g., the U-2 and SR-71 programs). The May 1960 shootdown of a U-2 over the Soviet Union gave great urgency to ongoing efforts to put photo reconnaissance satellites in orbit. The emerging Soviet nuclear threat now presented the possibility of extremely limited warning given the short flight times (i.e., under 30 minutes) for ICBMs and the much greater destructive potential of fusion weapons (which the Soviet Union first tested in 1953). The short-notice attack scenario would lead the United States to institute continuous airborne bomber alerts to supplement those on ground alert and also the maintenance of an airborne command post around the clock.[68]

---

[64] Kennedy's August 14, 1958, speech on the floor of the Senate drew on Alsop's earlier columns. See Christopher A. Preble, "Who Ever Believed in the 'Missile Gap'? John F. Kennedy and the Politics of National Security," *Presidential Studies Quarterly*, Vol. 33, No. 4, December 2003, pp. 801–826.

[65] Desmond Ball, *Politics and Force Levels: The Strategic Missile Program of the Kennedy Administration*, Berkeley, Calif.: University of California Press, 1980, pp. 50 and 57.

[66] Ball, 1980, p. 100. For more on the demise of the missile gap, see Ball 1980, pp. 88–104.

[67] Arthur M. Schlesinger, Jr., *A Thousand Days: John F. Kennedy in the White House*, New York: Mariner Books, 2002, p. 499.

[68] The airborne bomber alert program lasted from 1959 to 1967. During five of those years, between 10 and 12 aircraft were airborne at any given moment. During four of the years, between 3 and 9 aircraft were airborne. The airborne command post program kept one aircraft in the air 24/7 between its start in 1960 and its end in 1990.

Although the strategic nuclear competition continued to be a central element in the U.S.-Soviet competition, the Cold War became increasingly globalized, with both powers seeking points of leverage in every region. Some of the major crises and events of this period include the attempt by Cuban exiles to invade Cuba and overthrow Fidel Castro, which failed spectacularly at the Bay of Pigs (April 1961); the Berlin Crisis (June–November 1961); the Cuban Missile Crisis (October 1961); the growing U.S. military advisory role in Vietnam (1960–1964); the Soviet invasion of Czechoslovakia (1968); and major U.S. combat operations in Vietnam (1965–1972).[69]

The consequences of World War II were still being felt in the 1960s. The war had disrupted the global security environment in many ways, perhaps most powerfully by undermining the influence and control of the colonial powers over their prewar territories. This weakening led to a global decolonization movement, fueled by reborn nationalism and a desire for self-rule. Insurgencies became a growing challenge for colonial powers, and in some cases (e.g., Vietnam) the Soviet Union and China saw opportunities to advance their interests by supporting communist rebellions. The United States and its partners, in contrast, were generally on the defensive, seeing insurgencies as part of a Soviet-led global effort to replace Western-leaning governments with communist ones. As a result, American policymakers, academics, and analysts were increasingly preoccupied in the 1960s with the problem of rebellion and how it could be prevented through strategies that combined various civil and military instruments.[70]

In summary, the security environment of the 1960s was much more complicated than that of the 1950s, with strategic nuclear and conventional deterrence requirements, a growing war in Vietnam, and unrest in much of the developing world all vying for policymaker attention. Kennedy's strategy of "Flexible Response" was designed to better meet these new security challenges, offering a range of options tailored to the particulars of the conflict rather than the less flexible and nuclear-centric Massive Retaliation and New Look policies of Eisenhower.[71]

RAND research and analysis during the 1960s reflected these changes in the international security environment and new policies.

Although research on nuclear issues no longer dominated RAND's analytical agenda, important contributions continued to be made to deterrence theory and targeting concepts.

Bombers were also kept on ground alert between 1958 and 1990. For the years 1958 to 1967, between 223 and 625 bombers were on alert. Between 1968 and 1990, the numbers varied between 166 in 1968 and 53 in 1990. See Strategic Air Command, Office of the Historian, 1991, pp. 93 and 95.

[69] For an overview of American foreign policy during this period, see Steven W. Hook and John W. Spanier, *American Foreign Policy Since World War II*, Thousand Oaks, Calif.: CQ Press, 2018.

[70] For more on how the Cold War manifested in the developing world, see Robert J. McMahon, *The Cold War in the Third World*, Oxford, UK: Oxford University Press, 2013, and Odd Westad, *The Global Cold War: Third World Interventions and the Making of Our Times*, Cambridge, UK: Cambridge University Press, 2007.

[71] For more on Flexible Response, see Lawrence S. Kaplan, Ronald D. Landa, and Edward J. Drea, *History of the Office of the Secretary of Defense, Volume V, The McNamara Ascendancy: 1961–1965*, Washington, D.C.: Historical Office, Office of the Secretary of Defense, 2006, pp. 293–319, and Preble, 2004, pp. 3–6.

Technical assessments of targeting strategies had been part of RAND nuclear strategy research from the 1940s, seeking to answer such questions as "whether it was more efficient to attack concentrated industries or urban industrial complexes."[72] In the 1950s, the relative benefits of various targeting strategies were increasingly considered within the context of pre- and intrawar deterrence.[73] For example, Bernard Brodie argued for a no-cities war plan while working as a special consultant to the USAF Chief of Staff and brought these arguments against Massive Retaliation to RAND.[74]

The ideas being developed at RAND for more flexibility in targeting options were a perfect fit for the Kennedy's Flexible Response strategy, which sought to reduce dependence on strategic nuclear forces while expanding the range of nuclear options available to the President. As Soviet nuclear forces grew (albeit at a slower rate than had been feared) and their conventional forces expanded, Kennedy recognized that threats to massively retaliate in response to Soviet conventional aggression would become less credible. Thus, Flexible Response emphasized improvements in conventional military capabilities, particularly within the NATO alliance.

Within a month of Kennedy's inauguration, RAND analysts were briefing McNamara, the new Secretary of Defense, on these concepts. Most notable was William Kaufman's February 1961 briefing to McNamara on "counterforce" targeting. As described by Daniel Ellsberg, a RAND contemporary, Kaufman rejected plans for "all-out, nothing-held back" strikes against cities and industry and military targets, arguing "instead for developing a capability for sustained and controlled 'war fighting,' focused mainly on military targets, with cities withheld from initial attack."[75] Protecting U.S. bombers and missiles while attacking enemy nuclear forces was a key element of the counterforce concept, but the specifics of ABD/A were of less concern at this point than the interplay of strategies, escalation dynamics, and other higher order concerns.

Beyond nuclear issues, the new administration was increasingly concerned about lower level challenges, especially how to counter communist-inspired or supported insurgencies in the developing world. As the United States became more involved in the Vietnam conflict, RAND's research agenda increasingly focused on understanding the causes of insurgencies and identifying the most effective political and military policies to counter them. Although RAND

---

[72] Kaplan, 1983, p. 206.

[73] The relationship between war plans, targeting concepts, and military strategy remained a central and highly controversial aspect of nuclear policy throughout the Cold War. For more on these debates, see Scott D. Sagan, *Moving Targets: Nuclear Strategy and National Security*, Princeton, N.J.: Princeton University Press, 1990; David Alan Rosenberg, "The Origins of Overkill: Nuclear Weapons and American Strategy, 1945–1960," *International Security*, Vol. 7, No. 4, Spring 1983, pp. 3–71.

[74] Kaplan, 1983, p. 204.

[75] Ellsberg, 2017, p. 120. For more on the counterforce concept's development at RAND, see Ellsberg, 2017, pp. 119–128, and Kaplan, 1983, pp. 201–219. For more on the Kennedy administration's early thinking on nuclear strategy, see Scott D. Sagan, "SIOP-62: The Nuclear War Plan Briefing to President Kennedy," *International Security*, Vol. 12, No. 1, Summer 1987, pp. 22–51.

produced some well-known broader and more theoretical studies, such as the Leites and Wolf's *Rebellion and Authority*,[76] RAND's work on insurgency was overwhelmingly focused on helping its DoD clients win the war in Southeast Asia, producing at least 59 reports on that conflict between 1962 and 1969.[77]

Overview of RAND Research on Air Base Defense and Attack During This Period

The number of reports that RAND published on ABD/A dropped dramatically, from 61 in the 1950s to 30 in the 1960s. This decline reflected the shift toward other policy problems discussed above, as well as a growing confidence in the analytical and policy foundation regarding strategic nuclear issues (the driver of 1950s ABD/A research). This is not to imply that the problem was solved, or that policymakers were sanguine, just that as the knowledge base grew, the need for new analysis on topics such as nuclear effects was lessened. This also can be seen in the shift away from nuclear-related research. During the 1950s, RAND published 50 reports on nuclear threats to air bases; in the 1960s, RAND published only 11 dealing with nuclear issues, compared with 15 on conventional operations. The other major shift was to begin thinking about how to most effectively attack enemy bases. Whereas there had been no offensive-oriented reports in the 1950s, the 60s saw a more balanced analytical portfolio, with 10 reports on offensive options and 16 on defense.

By the 1960s, the USAF, although still attentive to nuclear threats, began to emphasize programs that would enhance defenses against conventional air attacks. Although a debate raged about the relative capabilities of the NATO and Warsaw Pact air forces, as well as the number of aircraft available for combat on both sides, few disputed that Soviet and Warsaw Pact air forces were sufficiently large and capable to successfully attack NATO air bases.[78] They would likely suffer significant attrition and be limited to daylight, clear-weather attacks using inaccurate dumb bombs. Nevertheless, aircraft in the open had proven to be extremely vulnerable to exactly such attacks during previous conflicts.

---

[76] Nathan Leites and Charles Wolf, Jr., *Rebellion and Authority: An Analytic Essay on Insurgent Conflicts*, Santa Monica, Calif.: RAND Corporation, R-462-ARPA, 1970. For an overview of RAND's work on countering insurgencies see Austin Long, *On "Other War:" Lessons from Five Decades of RAND Counterinsurgency Research*, Santa Monica, Calif.: RAND Corporation, MG-482-OSD, 2006.

[77] Austin Long identifies 19 of the major works on rebellion, insurgent motivation, and similar topics published during the 1960s (see Long, 2006, pp. 75–83). We did a quick search of RAND's reports published during the 1960s on the Vietnam War and found another 40 reports, for the total of 59. For more on RAND's role in the Vietnam War, see Mai Elliott, *RAND in Southeast Asia: A History of the Vietnam War Era*, Santa Monica, Calif.: RAND Corporation, CP-564-RC, 2010.

[78] For more on this debate, see Alain C. Enthoven and K. Wayne Smith, *How Much Is Enough? Shaping the Defense Program 1961–1969*, Santa Monica, Calif.: RAND Corporation, CB-403, 2005, pp. 142–147. (Originally published by Harper and Row Publishers in 1971.)

By 1961, NATO and U.S. leaders had decided to make every effort to reduce reliance on nuclear weapons through improvements in conventional forces.[79]

> Although Kennedy administration defense analysts stopped the program to build shelters hardened against nuclear attack, they favored the concept of "soft" shelters to protect parked aircraft against conventional munitions. . . . Secretary McNamara felt these shelters would be a more cost-effective means to improved USAF combat capability than investing in more unprotected aircraft.[80]

With the Soviet Air Force growing in size and capability, the threats to NATO bases in Europe grew as well. Because NATO counted on its air forces to offset the Soviet and Warsaw Pact advantage on the ground, it was imperative that NATO air bases be protected. RAND analysts Alain Enthoven and K. Wayne Smith noted that "Every pertinent study and war game conducted since 1961 has led to the same conclusion: actions taken to reduce the vulnerability of U.S. tactical air forces on the ground—particularly the building of shelters—will greatly increase our ability to fight a conventional war."[81] As Lawrence R. Benson noted in his 1981 study,

> A special USAFE staff study, published in September 1962, confirmed the command's vulnerability to conventional air attacks launched by the Warsaw Pact and recommended construction of soft aircraft shelters as well as improved point air defense, dispersal and camouflage. At an Air Staff conference later in the month, USAFE and PACAF submitted a requirement for 600 and 411 shelters respectively.[82]

To generate more momentum in this area, in September 1964 McNamara consolidated the various USAF and DoD efforts into the Theater Air Base Vulnerability Study (known as TAB VEE). The TAB VEE study (conducted by the USAF and DoD in 1964 and 1965 with substantial RAND support[83]) recommended dispersal, base hardening, and hardened shelters for fighter bases in Europe, Japan, and Korea.[84] These initiatives were vital, both to address the current threat but more importantly to be prepared for the great advances to come in the decade after Nikita Khrushchev was overthrown in 1964. One of the major histories of the Soviet Air Force observed

> During the Brezhnev regime, aircraft development proceeded steadily and the industry turned out a steady stream of ever-improved types. Although the world's

---

[79] Enthoven and Smith, 2005, p. 131.

[80] Lawrence R. Benson, *USAF Aircraft Basing in Europe, North Africa, and the Middle East, 1945–1980*, Ramstein Air Base, Germany: Headquarters, U.S. Air Forces in Europe, 1981, p. 113.

[81] Enthoven and Smith, 2005, p. 224. For more on the NATO-Warsaw balance during these years, see Richard A. Bitzinger, *Assessing the Conventional Balance in Europe, 1945–1975*, Santa Monica, Calif.: RAND Corporation, N-2859-FF/RC, May 1989; and Richard L. Kugler, *The Great Strategy Debate: NATO's Evolution in the 1960s*, Santa Monica, Calif.: RAND Corporation, N-3252-FF/RC, 1991.

[82] Benson, 1981, p. 114.

[83] Additional details regarding RAND's role in the TAB VEE study are provided later in this chapter.

[84] Benson, 1981, pp. 111–128.

attention was riveted on Soviet naval and missile developments in the late 1960s and 1970s, the growth in the numbers and capabilities of Soviet air power was just as spectacular.[85]

In the DoD Annual Report for fiscal year (FY) 1969 (written in January 1968), McNamara observed

> Over the past year, the great importance of adequate protection for air bases and aircraft in forward areas has again been dramatically demonstrated in the Middle East and in Southeast Asia. In a few hours of lightning strikes against the Arab's unprotected air bases and aircraft on 5 June, Israel annihilated the Arab air forces and achieved absolute air superiority in the combat zones for the duration of the six-day war.[86]

Regarding the 1967 Six-Day War, McNamara was referring to the surprise attacks by the Israeli Air Force against Egyptian, Syrian, Jordanian, and Iraqi airfields, which destroyed roughly 400 Arab aircraft and heavily damaged runways. These attacks showed how effective low-level strafing and bomb attacks by fighter-bombers could be against aircraft in the open. The only advanced weapon used by the Israeli's was a runway-busting bomb designed for low-altitude release.[87]

In response to what he saw as a growing fighter-bomber threat to U.S. air bases abroad, McNamara called for construction of new shelters. In his FY 1969 Defense Budget submission, McNamara wrote that USAF-designed fighter shelters had proven in tests to "provide excellent protection against anything but a direct hit by a conventional bomb, and some protection in a nuclear attack . . . together with the active defense by our CHAPARRAL and HAWK missiles and our VULCAN guns would provide a strong integrated defense for our overseas aircraft." The Secretary noted, however, that "while the Congress has appropriated funds for runway repair materials and equipment for various physical security measures, our past requests for aircraft shelter construction have been denied. . . . We are, therefore, again requesting funds . . . in FY 1969 . . . [that would] provide 60 shelters at European bases. As presently planned, the total program would provide shelter for 515 aircraft."[88]

Losses to North Vietnamese and Viet Cong mortar and rocket attacks on USAF bases in South Vietnam grew dramatically between 1965 and 1967, leading to an urgent program to build hardened shelters in that country.[89] In response to the rocket threat, 400 "Wonder" shelters were

---

[85] Kenneth R. Whiting, *Soviet Air Power*, New York: Routledge, 2019, p. 44. See also Mark O'Neill, "The Soviet Air Force, 1917–1991," in Robin Higham and Frederick W. Kagan, eds., *The Military History of the Soviet Union*, New York: Palgrave Macmillan, 2002.

[86] Robert S. McNamara, *Statement on the Fiscal Year 1969–73 Defense Program and the 1969 Defense Budget*, Washington, D.C.: U.S. Department of Defense, 1968, p. 137.

[87] John F. Kreis, *Air Warfare and Air Base Air Defense, 1914–1973*, Washington, D.C.: Office of Air Force History, 1988, pp. 315–319.

[88] McNamara, 1968, pp. 137–138.

[89] Vick, 1995, pp. 69–70.

built in Vietnam by USAF civil engineers between 1968 and 1970. By 1969, the first hardened fighter shelters were finally being constructed in Germany; a total of 1,000 would be built in Europe and Asia by the end of the Cold War.[90] Figure 4.1 illustrates a "Wonder" shelter located at Phu Cat Air Base, Vietnam.

Figure 4.1. F-4 Phantom Fighter in "Wonder" Shelter at Phu Cat Air Base, Vietnam

SOURCE: USAF photo provided by U.S. National Archives, File 342-KE-37286.

Table 4.1 lists the major policy emphases of RAND ABD/A research in the 1960s.

---

[90] Roger Fox, *Air Base Defense in the Republic of Vietnam, 1961–1973*, Washington, D.C.: Office of Air Force History, 1979, pp. 71–73; and Weitze, 2001, pp. 239–240.

Table 4.1. Top Policy/Analytical Objectives of RAND Reports on Air Base Defense and Attack: 1960–1969

Policy/Analytical Objective	Number of RAND Reports
Identify most effective means to damage enemy runways	9
Ensure strategic force survival	9
Conduct air operations during and after conventional attack	3
Conduct air operations during and after nuclear attack	3

NOTE: The table only includes reports in the top four categories, not all reports in the decade.

Out of the ten reports on offensive operations, nine were focused on how best to damage runways. The tenth report assessed the potential of conventionally armed intermediate-range ballistic missiles (IRBMs) to attack parked aircraft. Another nine reports continued the 1950s emphasis on ensuring the survival of strategic nuclear forces. Three reports assessed how best to conduct operations during and after a nuclear attack, and another three did the same for conventional strikes. Finally, six reports addressed the following objectives: (1) protect air bases from ground attack, (2) deter nuclear attack on air bases, (3) quantify effects of enemy conventional IRBM attacks on USAF bases in Vietnam (and nearby aircraft carriers), (4) defeat enemy air attacks, (5) protect air bases from enemy air attacks, and (6) identify most effective means to attack enemy airfields.

A Sampling of Reports

In this section, we will discuss four RAND reports from this period that reflect the breadth of research on ABD/A:

- *On Hardened Basing of B-52 Aircraft* (1961)
- *Tentative Thoughts on Non-Nuclear IRBM's for Attacking Parked Aircraft* (1963)
- *Airbase Defense and Security with Application to Thailand* (1966)
- *Some Hand-Done Calculations on Attacks Against Runways* (1969).

The first report, *On Hardened Basing of B-52 Aircraft*, continued the 1950s work on reducing the vulnerability of USAF strategic bombers to nuclear attack. It captures the technological optimism of the period, as well as the belief that strategic nuclear force survival was so vital that no idea should be ruled out as too wild or expensive. The authors discuss the technical feasibility and survivability of multiple hardened aircraft shelter options. They also offer some tentative thoughts on costs. The authors describe four options that they label "Conventional," "Cliff Dweller," "Mountain Side," and "Hardcore."

The Conventional shelter is an above-ground design that can be sized for alert, maintenance, or turn-around purposes. This option was the most advanced and understood in 1961. The authors evaluated several "rather detailed design studies . . . in order to determine suitable shelter configurations and the corresponding construction costs," concluding "that a covered arch is the most promising hardened hangar for the B-52 aircraft." The authors describe this earth-covered hangar (illustrated in Figure 4.2) as "essentially a large, reinforced concrete circular arch wide enough to admit the wing span and high enough to admit the tail of the aircraft," with the hangar floor "close to the grade of the taxiway or runway."[91]

Figure 4.2. RAND Design for B-52 Covered Arch Shelter

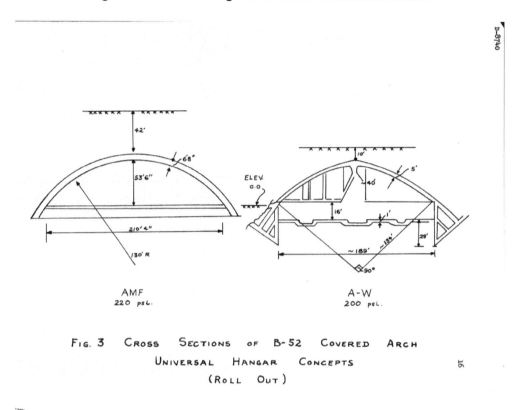

SOURCE: J. G. Hammer and Armas Laupa, *On Hardened Basing of B-52 Aircraft*, Santa Monica, Calif.: RAND Corporation, D-9513, 1961, p. 16.

Cliff Dweller was an extremely ambitious concept in the spirit of Herman Kahn's underground civil defense ideas (or Dr. Strangelove's, depending on one's perspective). Cliff Dweller sought to design an underground air base that could withstand nuclear attack. RAND teams published 15 reports in 1960 and 1961 focused on design, construction, vulnerability, and habitability of Cliff Dweller bases.[92] In terms of hardening, the Cliff Dweller was defined as

---

[91] J. G. Hammer and Armas Laupa, *On Hardened Basing of B-52 Aircraft*, Santa Monica, Calif.: RAND Corporation, D-9513, 1961, p. 6.

[92] See, for example, D. L. Lamar and D. Oberste-Lehn, *Operation Cliff Dweller: Determination of Site Location for Hardened Aircraft*, Santa Monica, Calif.: RAND Corporation, D-8078, 1960 (declassified by the Air Force Declassification Office on October 22, 2018); and A. R. Tamplin, *Operation Cliff Dweller: Hardening Bases, Atmospheric Control and Disease*, Santa Monica, Calif.: RAND Corporation, D-8020, 1960.

an aircraft base located under a mountain ridge. In its essential features, it consists of a central facility, an entrance tunnel for normal operations and several emergency exits. The central facility is placed under sufficient rock cover to make it virtually invulnerable to very impressive nuclear attacks. It houses all facilities required for aircraft storage and maintenance, personnel facilities, command and control facilities, weapon and POL storage as well as power and other household equipment.[93]

Mountain Side is defined as similar to mountain shelters already in existence in European countries, using "a short tunnel excavated into a steep rock face to house one or more aircraft."[94] Finally, the Hardcore concept is intended to make existing soft SAC bases capable of refueling and armed returning aircraft for additional strikes. The Hardcore base would not have shelters to protect aircraft from attack but would have survivable support facilities.

The authors conclude that Cliff Dweller would be too expensive and Mountain Side not feasible for B-52 aircraft. They recommend pursuit of the above-ground Conventional shelters as well as the Hardcore base recovery concept. As it turned out, the USAF rejected all of these hardening concepts, opting for less expensive alternatives to enhance survivability (e.g., improved early warning; airborne and ground alert for bombers and tankers).

The second report, *Tentative Thoughts on Non-Nuclear IRBM's for Attacking Parked Aircraft*, considers the potential of conventionally armed IRBMs as air base attack weapons. This possibility was explored at RAND as early as the 1950s, but missile accuracies were deemed inadequate. In this 1963 report, the RAND authors observe that "during the past decade the question of using ICBMs and IRBMs with non-nuclear warheads has been posed many times and in most cases has been discarded."[95] That report determined that if IRBM accuracies were improved to 1,500 ft circular error probable (CEP)—a measure of weapon system accuracy, in which a smaller number indicates greater precision—they could carry a sufficient number of submunitions (weighing 1.73 lbs each) to effectively attack aircraft parked in the open. Since the CEPs of the contemporary IRBMs (e.g., the Jupiter and Thor) were 4,800 and 7,100 ft, respectively, the authors concluded that airfield attack was not feasible "with non-nuclear ordnance using available ballistic missile designs."[96] By the mid-60s, however, some analysts believed that Soviet IRBMs (e.g., the SS-4) could be modified to carry large payloads of submunitions against targets under 1,000 km, trading range for payload. For example, a 1966

---

[93] Hammer and Laupa, 1961, p. 13.

[94] Hammer and Laupa, 1961, p. 15.

[95] B. F. Jaeger and M. B. Schaffer, *Tentative Thoughts on Non-Nuclear IRBMs for Attacking Parked Aircraft*, Santa Monica, Calif.: RAND Corporation, D(L)-11285-PR, 1963, p. 3.

[96] Jaeger and Schaffer, 1963, p. 12.

RAND study concluded that the SS-4 could achieve a CEP of 600 ft against targets within 1,000 km, with each missile delivering roughly 9,000 submunitions (weighing 0.63 lbs each).[97]

The third report, *Airbase Defense and Security with Application to Thailand*, tackles a problem that became more serious in the two years after it was published, the damage that Viet Cong and North Vietnamese Army mortars and rockets were doing to USAF aircraft based in Vietnam. Although only one report was published on ground threats to air bases, the report was interesting in several ways. It was RAND's first analysis of the topic, it presented a thoughtful application of lessons learned from ground attacks on air bases in Vietnam to the defense of USAF bases in the Thailand, and, surprisingly, it was sponsored by the Advanced Research Projects Agency rather than the USAF.[98] The report was published in December 1966 and thus was able to draw on roughly 19 months of experience in Vietnam (November 1964 to mid-1966).[99] The report offered a tactical and technical assessment of the threat, mitigation options, command and control of base defenses, personnel requirements for a ground defense infantry company, and how to develop an air base defense plan. The authors offer two major findings:

- Experience in South Vietnam has demonstrated that airbases are extremely vulnerable to ground attacks and that these attacks lean heavily on the enemy's ability to take advantage of surprise and concentration of forces.
- The analysis . . . suggests that the threat to airbases can be reduced by an energetic combination of active and passive defense and security measures including an efficient airborne alert team and ambush patrols working closely and in full cooperation with regional or local indigenous forces.[100]

The fourth report, *Some Hand-Done Calculations on Attacks Against Runways*, reflects the shift of RAND research from purely defensive considerations to analysis on offensive options against enemy air bases, in particular how best to damage enemy runways. The author, E. H. Sharkey, sought to supplement ongoing RAND computer simulations with some "hand-done" calculations, specifically looking at the "delivery of retarded weapons on high-speed low-altitude runs."[101] The author notes that earlier RAND work identified the effectiveness criterion such "that no 50-foot width of runway long enough for emergency takeoffs could be left for the

---

[97] J. G. Hammer and W. R. Elswick, *Conventional Missile Attacks Against Aircraft on Airfields and Aircraft Carriers*, Santa Monica, Calif.: RAND Corporation, RM-4718-PR, 1966. (Declassified by the USAF on October 23, 2018).

[98] The report may have been an analytical target of opportunity. The authors were both attached to ARPA's R&D Field Unit in Bangkok, Thailand, when they conducted the analysis.

[99] Robert Crawford and J. W. Ellis, Jr., *Airbase Defense and Security with Application to Thailand*, Santa Monica, Calif.: RAND Corporation, D-15350-ARPA/AGILE, 1966.

[100] Crawford and Ellis, 1966, pp. iii–iv.

[101] E. H. Sharkey, *Some Hand-Done Calculations on Attacks Against Runways*, Santa Monica, Calif.: RAND Corporation, D-18821-PR, 1969, p. 1.

enemy's use."[102] This was problematic because the experience during the Vietnam War with the F-111A indicated that the average bomb spacing was 100 ft, too big a gap between craters to close a runway. Concluding that we "seem to be stuck with the practical 100 foot spacing of retarded weapons," Sharkey explored whether low-level attacks at varying angles (0, 30, 45, and 90 degrees) to the runway might achieve the requirement "for hitting each 50-foot strip on a single pass."[103] In a series of charts displaying area-hit probabilities, single-pass probabilities of hits, and the expected number of hits for four different CEP values, Sharkey quantifies the trade-offs facing the tactical commander, including "whether the commander wants the largest number of hits on the runway for every ten missions (say), or whether he would prefer to almost guarantee that the runway will be hit on every pass."[104] The Sharkey paper is representative of much work at RAND in the 1950s and 1960s in which hand calculations and hand-drawn graphs were used to productively explore aspects of complex operational problems and supplement the results from computer simulations (*machine runs*, in the parlance of the day).

RAND Contributions

As noted earlier, RAND devoted considerable effort to supporting DoD clients during the Vietnam War. Despite the demands of that work, RAND analysts continued to make new and significant contributions to ABD/A. Table 4.2 highlights four of these contributions.

First, RAND efforts to design and assess alternative concepts for protecting SAC bombers from nuclear attack continued into the early 1960s. By 1964, these efforts had expanded to include how to protect fighter aircraft from conventional attack. In addition to its own studies on shelter design, RAND provided analytical support to a major USAF initiative, the Theater Air Base Vulnerability Study Group. The study included operational analyses, construction designs, and field exercises. The name of the major field exercise, Theater Air Base Vulnerability Evaluation Exercise, was abbreviated to TAB VEE. Both the larger study and the hardened aircraft shelters that were ultimately built (beginning in the late 1960s and continuing in the 1980s) in Europe and Asia became known as TAB VEE's.[105] The original Wonder shelter design was improved through the addition of hardened doors and a full back wall, exhaust ports, and other features. It also was enlarged to accommodate larger aircraft (e.g., the F-15) that were coming in the force in the 1970s. Figure 4.3 illustrates the "2nd Generation HAS" (hardened air structure) version of the final design found at Kadena Air Base Japan, with hardened doors open to the sides.

---

[102] Sharkey, 1969, p. 1.

[103] Sharkey, 1969, p. 3.

[104] Sharkey, 1969, p. 3.

[105] See Weitze, 2001, p. 235, for more on the TAB VEE effort.

Figure 4.3. F-15 in Hardened Aircraft Shelter, Kadena Air Base, Japan

SOURCE: Dani Johnson, "Kadena Prepares for Typhoon," U.S. Air Force, July 12, 2007.

The RAND team was recognized in the report preface for their contributions as follows: T. E. Greene (analysis, gaming, models), Robert Keese (ATTRIT model), Robert Martin (ADVAL AAGUN model) and Natalie Wilson[106] (ADVAL model). ATTRIT was used to simulate the air battle and attrition of aircraft on airfields. ADVAL (Active Defense Evaluation Model) assessed the performance of active defenses in airfield defense. Computer simulations and models were still considered novel and the TAB VEE report even specifies that ATTRIT was run on a RAND IBM 7044 computer, while ADVAL was originally written for the RAND JOSS computer system.[107] (We'll say a bit more about JOSS in Chapter 5.)

Second, RAND began to study how to best attack enemy air bases using conventional weapons, with an initial focus on closing runways. Although RAND analysts did not design the runway cratering device recommended by Thomas Edwards (1966), they were intimately involved in the test and evaluation process, helping to solve several critical design problems. In Edwards' words "[RAND analysts] Marv Schaffer, Bernie Jaeger, and Jack Ellis were able to

---

[106] RAND analyst Natalie Wilson is better known by her married name: Natalie W. Crawford.

[107] The TAB VEE study report is not publicly available. References to RAND can be found in Volume One as follows: ATTRIT model (pp. 6, 3, 298) and reference to RAND IBM 7044 (p. 298). References to RAND in Volume Two can be found as follows: ATTRIT model (p. 397), ADVAL and JOSS (pp. 442, 600).

recognize the principles of mechanics which were represented by the bizarre movements of the linked components of the parachute-weapon system and to suggest efficient methods for putting the several hypotheses to test."[108]

Third, RAND analysts argued in 1963 that IRBMs were now sufficiently accurate that, if armed with hundreds of small submunitions, they could be effective weapons for attacking enemy aircraft parked in the open. Similarly, a different RAND team, writing in 1966, concluded that if the Soviet Union provided a submunition-armed IRBM to the North Vietnamese, then USAF bases in South Vietnam and aircraft carriers operating nearby could be attacked in a similar manner. The United States chose not to pursue this technology, and thankfully IRBM attacks on air bases in Vietnam never transpired. This work was prescient in analytically demonstrating the potential value (and threat) of submunition-armed ballistic missiles. We will return to this topic in Chapter 6 (the 1990s), when RAND analysts reconsidered this threat in light of the revolution in accuracy brought by Global Positioning System (GPS) guidance systems.

Finally, Crawford and Ellis were the first to assess how the air base attack tactics used by the Viet Cong and North Vietnamese Army in Vietnam might be applied more broadly, specifically against USAF bases in Thailand. Ground attacks on U.S. air bases in Vietnam ultimately destroyed 393 U.S. and allied aircraft and damaged another 1,185, demonstrating the seriousness of this threat.[109] RAND would return to this topic in the mid 1990s, considering the attractiveness of air base ground attack as an adversary option to counter American air power.

Table 4.2. Major RAND Contributions During the 1960s

RAND Contribution	Documented in
Engineering analysis of hardened aircraft shelter designs	Hammer (1961), Hammer and Laupa (1961), Hammer and Sandoval (1961)
Analysis of runway attack tactics and weapons choices	Green (1963), Wilson and Jaeger (1966), Edwards (1966), Sharkey (1969)
Viability of conventionally-armed IRBMs in air base attack	Jaeger and Schaffer (1963), Hammer and Elswick (1966)
Application of Vietnam lessons learned to air base ground defense in Thailand	Crawford and Ellis (1966)

The next chapter discusses RAND ABD/A research in the 1970s and 1980s.

---

[108] T. I. Edwards, *Successful Tests of a RAND-Recommended Runway Cratering Device*, Santa Monica, Calif.: RAND Corporation, D-15075, 1966, p. 6.

[109] Alan Vick, *Snakes in the Eagle's Nest: A History of Ground Attacks on Air Bases*, Santa Monica, Calif.: RAND Corporation, MR-553-AF, 1995, p. 19.

5. Conventional Warfare in Central Europe, 1970–1989

Strategic Environment

By the start of the 1970s, the United States was beginning to draw down its forces in Vietnam. To facilitate its exit from the war, the U.S. military adopted a policy of "Vietnamization," whereby it trained and equipped South Vietnamese forces so they could assume responsibility for their own defense. After reaching a peak of 549,000 U.S. troops in Vietnam in 1969, only 69,000 remained by 1972.[110] On January 26, 1973, the Paris Peace Accords were signed, marking the end of direct U.S. military involvement in Vietnam.

Meanwhile, the Cold War pressed on. As Soviet conventional and nuclear capabilities grew, there were concerns in Europe that the United States commitment to NATO might waver in the face of threats to the American homeland. One observer at the time cleverly noted that the United States "would not risk New York to save Paris." In response to these pressures, as well as the limitations of the earlier Massive Retaliation strategy, American military strategy had shifted to Flexible Response in the early 1960s, which was then adopted by NATO in 1967 and remained NATO doctrine well into the 1980s.[111] The doctrine was grounded in a mutually reinforcing posture of conventional and nuclear deterrence. NATO's conventional forces were forward deployed along its eastern border in an effort to deter a Pact offensive. If conventional deterrence failed, Flexible Response dictated that NATO would employ theater nuclear weapons defensively to deescalate tensions. And if that failed to restore deterrence, massive retaliation with strategic nuclear forces would be the final resort.[112]

At around this same time, anti–ballistic missile (ABM) systems were also in development. ABM systems capable of intercepting nuclear-tipped ballistic missiles could theoretically deny or severely degrade a nuclear strike if deployed in large enough quantities. Proponents of ABM defenses saw their main utility as a way to defeat nuclear attack from nations with a relatively small nuclear arsenal, especially China, which was viewed by some strategists as irrational and perhaps not deterrable. However, ABM systems also threatened to undermine the strategic stability that Flexible Response sought to achieve with the Soviet Union. If one of the superpowers could deploy effective ABM systems in large numbers, it might be able to

---

[110] Office of the Federal Register, National Archives and Records Administration, "News Conference of Secretary of Defense Melvin R. Laird Following the President's Announcement," January 13, 1972, in *Weekly Compilation of Presidential Documents*, Quarterly Index, First Quarter, January–March 1972.

[111] NATO's Military Council Document 14/3 described Flexible Response as "forward defense with flexibility." See J. Michael Legge, *Theater Nuclear Weapons and the NATO Strategy of Flexible Response*, Santa Monica, Calif.: RAND Corporation, R-2964-FF, 1983, p. 1.

[112] Legge, 1983, and Steven L. Canby, *NATO Military Policy: Obtaining Conventional Comparability with the Warsaw Pact*, Santa Monica, Calif: RAND Corporation, R-1088-ARPA, 1973.

challenge or call into question the legitimacy of its nuclear deterrent.[113] Research was underway in the United States to modify the existing Nike air defense system into a system capable of defending against ballistic missile threats. These efforts led to work on the Nike Zeus and Nike-X ABM programs in the late 1960s. The Soviet Union had its own ABM counterparts, most notably the A-35.[114]

The United States intended to deploy these systems in moderate numbers in critical locations across the country, but those plans never came to full fruition. The cost and limited effectiveness of ABM systems ultimately made them unworkable as an alternative to deterrence. For one, their effectiveness was severely limited by the concurrent development of ICBM payloads, which could split into numerous warheads prior to reentry—before they could be struck by existing and planned ABM systems. The development of these multiple reentry vehicles (MRVs) and multiple independently targetable reentry vehicles (MIRVs) meant that ABM interceptors would need to be available in sufficient numbers to target every one of the MIRVs to effectively negate a single ICBM. Aside from being a difficult technological feat, it was unfavorable from a cost-exchange perspective to shoot down MIRVs. It was simply too expensive to field the number of ABMs (typically at least two interceptors per target) necessary to destroy all of them.[115] Thus, deploying ABM systems in large numbers was impractical from an economic standpoint.

The first round of Strategic Arms Limitation Talks (SALT) in 1969 and the agreements that followed in 1972 officially settled the matter of ABM systems. In addition to freezing the overall size of each country's arsenal of ballistic missiles and launchers, SALT I led to the Anti-Ballistic Missile Treaty, which limited the deployment of ABM systems to just two sites in each country.[116] These arms control measures ushered in a détente between the superpowers for most of the decade.

However, even during this period of somewhat relaxed tensions (and in the years that followed), there was an ongoing public and private debate about the credibility of NATO's conventional deterrent. Many experts argued that NATO's relatively small conventional force was spread too thin along its eastern border and was vulnerable to a Warsaw Pact offensive. The most likely scenario, according to Soviet warfighting doctrine, was a blitzkrieg-style attack across Central Europe in which the Warsaw Pact would concentrate forces at certain locations, achieve an overwhelming superiority in force ratios, and penetrate deeply with armored columns

[113] J. I. Coffey, "The Anti-Ballistic Missile Debate," *Foreign Affairs*, Vol. 45, No. 3, April 1967, pp. 403–413.

[114] Alexander Flax, "Ballistic Missile Defense: Concepts and History," *Daedalus*, Vol. 114, No. 2, Weapons in Space, Vol. I: Concepts and Technologies, Spring 1985, pp. 33–52.

[115] Flax, 1985; and U.S. Department of State, Treaty Between the United States of America and The Union of Soviet Socialist Republics on The Limitation of Anti-Ballistic Missile Systems (ABM Treaty), signed May 26, 1972.

[116] Flax, 1985.

into NATO territory along designated axes of advance.[117] The strategy involved an echelonment of forces in which succeeding waves of troops would apply continuous pressure on single points until they eventually broke through, at which point they could maneuver rapidly toward NATO's rear. Once deep in NATO territory, Pact forces could sever critical lines of communication and disable NATO's C2 infrastructure, thus compelling its surrender.[118]

Whether the swift march to victory would be as simple in reality as it sounded on paper was a topic of debate.[119] A numerical advantage in force structure is of course just one predictor of an outcome in war.[120] And while the Warsaw Pact had a much greater ratio of tanks and divisional manpower, there was a convincing argument that NATO would fare better in a conventional conflict than the prevailing opinion would indicate. Assuming that NATO forces would be spread evenly on its border and have reasonable indications and warning of Pact mobilization, the Pact's options would be limited to just a few axes of advance. These axes would be fairly predictable, and rapidly traversing Central Europe would be difficult because of obstacles in the terrain.[121] A slowdown in the blitzkrieg would at least provide NATO's political and military leaders time to coordinate their use of theater nuclear weapons, if necessary.[122] Moreover, if it became a war of attrition, the advantage would likely shift in favor of the army with the more robust economy.[123]

The merits of these arguments notwithstanding, NATO's ability to counter a Warsaw Pact offensive needed to be convincing not just to Western audiences but also to Soviet leadership. Critics argued that in order for Flexible Response to be effective writ large, NATO's conventional deterrent needed to be rock solid. This was because there were serious doubts at the time that the United States would honor its nuclear guarantee to Europe and put its own survival at risk in the event that NATO's conventional defense failed. By the mid-1970s, the U.S. and Soviet arsenals of theater nuclear weapons were similar in size, and, perhaps more importantly,

---

[117] For contemporary treatments of the military balance in Europe, see William Mako, *U.S. Ground Forces and the Defense of Central Europe*, Washington, D.C.: Brookings Institution, 1983; and Robert Shishko, *The European Conventional Balance: A Primer*, Santa Monica, Calif.: RAND Corporation, P-6707, 1981.

[118] James C. Barbara and Robert F. Brown, "Deep Thrust on the Extended Battlefield," *Military Review*, October 1982, p. 22; and John Woodmansee, "Blitzkrieg and the AirLand Battle," *Military Review*, August 1984.

[119] See, for example, Phillip A. Karber, "In Defense of Forward Defense," *Armed Forces Journal*, May 1984, p. 28.

[120] There was considerable debate at this time about analytical methods and metrics. See, for example, J. Stockfisch, *Models, Data, and War: A Critique of the Study of Conventional Forces*, Santa Monica, Calif.: RAND Corporation, R-1526-PR, 1975.

[121] John J. Mearsheimer, "Why the Soviets Can't Win Quickly in Central Europe," *International Security*, Vol. 7, No. 1, Summer 1982, pp. 3–39.

[122] There was a lively debate among defense analysts regarding the feasibility of blitzkrieg in the modern era, particularly between Jeffrey Record and Colin Gray. See Jeffrey Record, "The October War: Burying the Blitzkrieg," *Military Review*, April 1976; and Colin S. Gray, "The Blitzkrieg: A Premature Burial?" *Military Review*, October 1976.

[123] Mearsheimer, 1982.

Soviet doctrine stated that it would respond to NATO first use of theater nuclear weapons with massive retaliation. Detractors of Flexible Response argued that deterrence by punishment was only effective when it was possible to make the original offender "worse off in both relative and absolute terms." If this was impossible, would the United States risk its own survival for the sake of Europe?[124]

These doubts were amplified by voices in the expert community who strongly opposed what they viewed as an overreliance on nuclear weapons. Many advocated a "no first use" policy that would only permit NATO to launch a nuclear strike in response to an adversarial nuclear strike.[125] Those in the "no first use" camp recommended bolstering NATO's conventional capabilities and discarding the idea of a nuclear threshold altogether. However, Soviet leadership may have looked fondly on NATO adopting such a policy, since they perceived Warsaw Pact conventional strength to be superior. Those in the West who opposed no first use argued that NATO's relative conventional strength was unlikely to drastically improve, and therefore nuclear weapons were all the more important; no first use would only be reasonable if NATO's conventional strength was above reproach. Even those who believed NATO's conventional disadvantage was overblown did not believe that conventional defense was as compelling as nuclear defense.[126] The advantage of NATO's nuclear deterrent (if credible) was that it offered the twofold benefit of denial and retaliation, whereas its conventional threat only offered denial. Ultimately, this encapsulated the strategic predicament of the time: NATO's conventional deterrent was credible but not highly capable, whereas its nuclear deterrent was capable but not highly credible.[127]

Independent of the "no first use" debate, there was a consensus in the defense community that NATO needed to strengthen its conventional forces, especially as tensions escalated at the start of the 1980s. There were a number of ways for NATO to accomplish this. In theory, it could drastically boost defense spending on traditional weapon technology to more closely match Warsaw Pact force sizes, but such a large investment was not politically viable.[128]

One alternative was to adopt newer and more sophisticated weapons that could paralyze a Soviet blitzkrieg.[129] In aggregate, the breakthrough in Western military technology developments

---

[124] Legge, 1983.

[125] For discussions of nuclear escalation issues, see Fen Osler Hampson, "Groping for Technical Panaceas: The European Conventional Balance and Nuclear Stability," *International Security*, Winter 1983/84.

[126] John J. Mearsheimer, "Nuclear Weapons and Deterrence in Europe," *International Security*, Vol. 9, No. 3, Winter 1984–1985.

[127] Samuel P. Huntington, "Conventional Deterrence and Conventional Retaliation in Europe," *International Security*, Vol. 8, No. 3, Winter 1983–1984.

[128] Huntington, 1983–1984.

[129] For a 1970s view of these technologies, see James Digby, *Precision-Guided Munitions, Adelphi Paper 118*, London: International Institute for Strategic Studies, 1975. For a 1980s perspective on the contribution of PGMs to

from the mid 1970s to late 1980s became known as the second "Offset Strategy," since it helped alleviate NATO's numerical disadvantage just as nuclear weapons had after World War II.[130] The most critical technologies attributed to this period were probably improvements in targeting (better sensors, data integration, and C2), low observable aircraft, GPS, and an assortment of precision-guided munitions (PGMs).[131]

To successfully exploit these new weapons systems, U.S. Army leadership was concurrently spearheading the design of a new land warfare doctrine. Its extant doctrine of Active Defense (adopted in 1976 and aligned with NATO's doctrine of Forward Defense) called for shifting forces laterally in reaction to a Pact offensive, at which point Army fires would concentrate on specific "killing zones" as Pact forces advanced sequentially into them. But, as discussed, U.S. forces would be vulnerable to an overwhelming Pact blitzkrieg under this doctrine, especially if a controlled retreat was impossible. Critics pointed to an overreliance on firepower over maneuver, and attrition at the expense of initiative and tactical creativity.[132]

Released in 1982, the AirLand Battle doctrine sought to rebalance those aspects of Active Defense and effectively use the new resources at the U.S. military's disposal. Advanced sensors and weaponry provided U.S. forces with the ability to track and engage the enemy from great distances. They could seize the initiative and operate at depth instead of waiting for the next Soviet divisional wave to enter a narrow killing zone. In its application to Europe, AirLand Battle advocated targeting the Warsaw Pact's second echelon in the "deep battle" in order to isolate the first echelon and prevent the Pact from gaining an overwhelming force ratio in the close-in battle.[133] This idea of an extended battlefield was the central theme of the doctrine.[134] Although AirLand Battle was U.S. Army doctrine and not NATO doctrine, it was influential inasmuch as it offered a unified theory for the application of new weapon systems and helped restore credibility to U.S. conventional power.[135]

---

deterrence, see John J. Mearsheimer, *Conventional Deterrence*, Ithaca, N.Y.: Cornell University Press, 1983, especially chapters six and seven.

[130] For more on the second Offset Strategy, see Robert R. Tomes, *U.S. Defense Strategy from Vietnam to Operation Iraqi Freedom: Military Innovation and the New American Way of War, 1973–2003*, New York: Routlege, 2007.

[131] Advances in the capabilities of U.S. and NATO air power were central to this strategy. For an elegant treatment of the operational significance of these improvements, see Benjamin S. Lambeth, *The Transformation of American Air Power*, Ithaca, N.Y.: Cornell University Press, 2000.

[132] John L. Romjue, *From Active Defense to AirLand Battle: The Development of Army Doctrine 1973–1982*, Fort Leavenworth, Kans.: U.S. Army Training and Doctrine Command, TRADOC Historical Monograph Series, June 1984.

[133] For more on USAF-Army cooperation in the deep battle, see Lambeth, 2000, pp. 83–91.

[134] See Donn A. Starry, "Extending the Battlefield," *Military Review*, March 1981; and John Romjue, "The Evolution of the AirLand Battle Concept," *Air University Review*, May/June 1983, pp. 7–8.

[135] Although ALB was not the only deep attack concept, it arguably drove a robust and high-level debate among military officers and defense professional in the United States and Europe. See, for example, Boyd Sutton, John R. Landry, Malcolm B. Armstrong, Howell M. Estes, and Wesley K. Clark, "Strategic and Doctrinal Implications of Deep Attack Concepts for the Defense of Central Europe," in Keith A. Dunn and William O. Staudenmaier, eds.,

The 1980s concluded triumphantly for the United States. The Soviet Union was in a poor negotiating position because of its struggling economy and failed campaign in Afghanistan. Under Mikhail Gorbachev's leadership, the Intermediate-Range Nuclear Forces (INF) Treaty, which had previously seemed unattainable, was signed in 1987. The United States successfully avoided a hot war with the Soviet Union as it entered the new decade.

Overview of RAND Research on Air Base Defense and Attack During This Period

After a slight dip in the production of RAND reports on ABD/A in the 1960s, there was a resurgence in the number of reports on the subject in the 1970s and 1980s. Exactly 53 reports on ABD/A were produced in each of the decades, just shy of the 61 reports produced in the 1950s.[136] A likely explanation for the lower report count in the 1960s was the Vietnam War's outsized influence on the RAND research agenda at the time. Roughly 60 RAND reports were produced during the 1960s on insurgency, with 40 of those on the Vietnam War. If the Vietnam War had not happened, it is unlikely that more than a handful of those reports would have been written, and significant RAND research manpower would have been available for other activities.[137]

Here, we treat the 1970s and 1980s as a single chapter rather than two separate chapters because the strategic environment and key research themes during this time period were inextricably connected and overall quite similar. It should come as no shock that reports on ABD/A during each decade were dominated by research and analysis on the European theater. More than 75 percent of reports in the 1970s were focused on Europe; this number increased to more than 85 percent in the 1980s. Similarly, the reports in this time period were overwhelmingly related to conventional operations, with more than 80 percent and 85 percent of reports on this topic in each decade, respectively (see Chapter 2).

In contrast, nuclear issues were only discussed in seven reports over the whole 20-year span (see Chapter 2). This low number might be surprising given that air bases were high-value targets for theater nuclear weapons, which were so important under the Flexible Response strategy. One explanation is that RAND and other research organizations had so thoroughly investigated the effects of tactical nuclear weapons on air bases in the past that the problem was

---

Military Strategy in Transition: Defense and Deterrence in the 1980s, Carlisle Barracks, Pa.: U.S. Army War College, 1984.

[136] This overall count includes reports such as the TSAR and TSARINA manuals that had multiple volumes and versions.

[137] Although we cannot know how the actual research agenda would have developed, it is not unreasonable to postulate that the number of reports on ABD/A might have remained fairly flat throughout the Cold War. If we assume that half of effort devoted to Vietnam would have remained on ABD/A, the 1960s may have been similar to the following two decades in overall report count.

already well understood to the USAF. Moreover, as one RAND report pointed out, certain analytical methods that were popular in the 1970s and 1980s were more appropriate for conventional warfare than nuclear issues. Analyzing the cost-benefit of a narrowly defined conventional exchange was easier than trying to quantify the huge civil losses and socio-political upheaval that were likely in a nuclear exchange. Underscoring this point, the report states, "It is easier to consider how we feel about losing a few aircraft in exchange for a few of the enemy's than to think about Armageddon, and we should thus be better able to construct the value function required if the situation is relatively limited and free of imponderables."[138] In summary, conventional problems were overall more appropriate and tractable to address using the latest modeling tools at RAND's disposal.

While the 1970s and 1980s shared many similarities, it is also noteworthy that 1970–1979 was unique for being the only decade to have more reports on air base attack than air base defense. Many of these reports were dedicated to technical examinations of narrow engineering topics, such as aircraft payload sizes, munitions effectiveness, and shelter designs. However, others focused on novel targeting concepts, concepts of operation for standoff weapons, and cost-effectiveness analysis. The shift in focus toward attacking adversary air bases was likely influenced by NATO's need to negate Warsaw Pact airpower should a conventional war break out on the continent. As discussed in the previous section, the prevailing Warsaw Pact strategy was to initiate a blitzkrieg attack that would swiftly penetrate NATO's forward defenses and sever critical C2 nodes in NATO's rear. Close air support would be crucial to the success of such an attack, given its speed and flexibility. John Mearsheimer in *Conventional Deterrence* explains the benefit of having close air support over land-based artillery in a blitzkrieg:

> Dependence on land-based artillery presents two major problems. First, artillery exchanges waste valuable time, although the amount of time expended depends on whether the offense uses towed or self-propelled artillery. Second, extensive use of artillery requires large increases in the logistical support to sustain an attack. As the mass of the attacking force increases in size, the velocity naturally decreases. Given the critical importance of timing for a blitzkrieg, reductions in speed imperil its very existence. Close air support, on the other hand, presents none of these problems. Because it is inherently flexible, the airplane functions as the perfect complement to fast-moving armored forces.[139]

Thus, investigating better methods of attacking Warsaw Pact air bases to deny airpower was of great interest to U.S. policymakers at the time.

In the 1980s, there was a major shift back to base protection measures; 70 percent of reports were exclusively about defense (see Chapter 2). While dispersal, hardening, and runway recovery were the most common passive defense options discussed in the literature, they did not

---

[138] Carl Richard Neu, *Attacking Hardened Air Bases (AHAB): A Decision Analysis Aid for the Tactical Commander*, Santa Monica Calif: RAND Corporation, R-1422-PR, 1974.

[139] Mearsheimer, 1983, p. 40.

dominate as much as previous decades. On the contrary, what stood out about this time period was the wide variety of passive and active defense options that were explored. Exploring the effectiveness of different passive defense options was easier now than ever before because of RAND's improved modeling capabilities, as discussed in the section on "RAND Contributions" later in this chapter.

Supplementing the statistics referenced from Chapter 2, Table 5.1 summarizes the top policy and analytical objectives explored during these two decades.

Table 5.1. Top Policy/Analytical Objectives of RAND Reports on Air Base Defense and Attack: 1970–1989

Policy/Analytical Objective	Number of RAND Reports
Identify most effective means to attack enemy airfields	29
Conduct air operations during and after conventional attack	24
Identify most effective means to defend airfields	17
Identify most effective means to attack enemy hardened aircraft shelters	12
Ensure strategic force survival	5

NOTE: The table only includes reports in the top five categories, not all reports in the two decades.

A Sampling of Reports

In this section, we will discuss three RAND reports from this period that capture a mix of research topics on ABD/A:

- *Target-Marking Systems for RPVs Used as Designator Vehicles for Airbase Attack* (1971)
- *An Approach to Studying Methods of Achieving Air Superiority by Attacking Enemy Airfields* (1974)
- *Tactical Dispersal of Fighter Aircraft: Risk, Uncertainty, and Policy Recommendations* (1987).

The first report, published in June 1971, was a RAND internal note titled *Target-Marking Systems for RPVs Used as Designator Vehicles for Airbase Attack*. This document (along with a companion piece in the same month) was an early examination of a new concept of operation for remotely piloted vehicles (RPVs). RAND analyst William H. Krase envisioned conducting hunter-killer operations at adversary air bases using RPVs to feed targets to incoming attack vehicles. The report investigates the necessary conditions for the RPV to perform target identification and designation while remaining survivable. Krase explored two main challenges: (1) identifying Soviet hangarettes using infrared imagery and (2) minimizing the risk to survivability while using a reliable target marking technique. The report concluded that "the most feasible designator concept involves night operation, a FLIR [forward-looking infrared] sensor, and a laser-marking device operating through the same stabilized and widely slewable optics. The concept implies fairly long (e.g. 5 min) exposure of a low-altitude, low-speed

49

designator vehicle for each weapon delivery." While not comprehensive or definitive in its findings, this report was one of many others in this period that analyzed the feasibility of unconventional air base attack strategies and in particular the use of RPVs.[140]

The second report, *An Approach to Studying Methods of Achieving Air Superiority by Attacking Enemy Airfields*, was published in 1974. This report examined options to decrease adversary sortie generation under surge conditions. The author, Sidney H. Miller, first points out actions that can temporarily increase sortie generation and produce surge capacity. These include deferring maintenance, performing maintenance concurrently, flying aircraft immediately once they are ready, working more hours at greater productivity, giving long maintenance jobs low priority, cannibalizing parts from other aircraft, operating from forward bases, and reducing or eliminating delay times. Next, Miller breaks down ways to decrease adversary sorties. These fall into five broad categories: (1) decreasing the number of enemy aircraft, (2) decreasing the mean length of enemy flying day, (3) increasing the mean length of enemy sortie cycle, (4) increasing enemy maintenance time, and (5) targeting parts of a base that increases maintenance.

Miller's treatment of the final two categories related to maintenance is especially detailed and thought-provoking. Miller notes that air missions that are more difficult for enemy aircraft— for example, missions that involve use of the afterburner, hard field landings, or combat damage—increase the probability that the aircraft will require maintenance after a sortie.[141] He offers numerous maintenance-related targets, such as communications equipment, transportation equipment, aerospace ground equipment (AGE), buildings and docks, fuel support equipment, and aircraft mechanics, and emphasizes that future air base attack models should consider all of these factors. Miller wrote this report in 1974, and thus it should be seen as a precursor to the work of Donald Emerson, who built the TSAR (Theater Simulation of Airbase Resources) model about eight years later, which included many of the above points, particularly those related to maintenance and theater logistics.[142]

The next report, *Tactical Dispersal of Fighter Aircraft: Risk, Uncertainty, and Policy Recommendations*, by RAND analyst John Halliday, was published in 1987. The impetus for this report was the growing consensus regarding three fundamental facts: (1) NATO ground forces were dependent on friendly air forces to protect them from Warsaw Pact air attacks and to counterbalance the Warsaw Pact ground force advantage, (2) the ability of NATO air forces to generate the necessary sorties was entirely contingent on NATO air force main operating base (MOB) survivability, and (3) MOBs in Europe were increasingly vulnerable to attack by Warsaw Pact air forces and tactical ballistic missiles.

---

[140] W. H. Krase, *Target-Marking Systems for RPVs Used as Designator Vehicles for Airbase Attack*, Santa Monica, Calif: RAND Corporation, IN-21646-PR, 1971.

[141] Sidney H. Miller, *An Approach to Studying Methods of Achieving Air Superiority by Attacking Enemy Airfields*, Santa Monica, Calif: RAND Corporation, 1974, not available to the general public.

[142] Donald E. Emerson, *An Introduction to the TSAR Simulation Program: Model Features and Logic*, Santa Monica, Calif.: RAND Corporation, R-2584-AF, 1982.

The report attributes the consolidation of NATO air forces on a relatively small number of MOBs to five different factors: (1) a historical presumption of air superiority and tactical aircraft safety, (2) a lack of recognition that sustained sortie rates would be needed under the Flexible Response doctrine, (3) a bias toward using the economies of scale afforded by large bases to save money, (4) a preference for geographical proximity to facilitate command and control hierarchy, and (5) a tendency to leave support equipment on the ground instead of on the aircraft in an attempt to achieve better performance characteristics.

The study made a unique contribution to air base resiliency analysis by breaking down the different sources of uncertainty in air base operations (for example, related to detection, penetration, weapon delivery, and number of aircraft on base at a given time) into separate probability distributions and predicted a range of outcomes based on those uncertainties. The study examined the effect of dispersing a fighter wing to four separate dispersed operating locations. The analysis concluded that this dispersal method provided modest improvements in sortie generation and reductions in aircraft lost, but Halliday's strongest recommendation was that USAF leadership initiate a rigorous testing regime for defense capabilities, such as the effects of modern munitions on shelters and the capacity for rapid runway repair. He argued that new testing would help to refine many modeling assumptions made in this area.

RAND Contributions

Many reports on air base attack in the early 1970s were technical analyses of narrow subjects, such as the effect of various munitions on Soviet hangarette designs. As in the 1950s and 1960s, these technical investigations often included calculations and charts created painstakingly by hand. However, starting in the 1960s, computing power at RAND began to increase and open the door for more complex calculations. RAND developed the interactive and user-friendly computing environment called JOSS in 1963.[143] Well over a dozen USAF facilities across the country adopted JOSS in the following years, and by 1970 there were 500–600 RAND researchers and USAF personnel who could use JOSS terminals for their computing needs.[144] JOSS was eventually phased out, but modeling and simulation for defense purposes remained a staple of RAND research.[145] Modeling and simulation of ABD/A dynamics flourished from the

---

[143] Of course, *user-friendly* is a relative term. JOSS was a big improvement, but few 21st century computer users would find it or any other 1960s computer operating environment user-friendly.

[144] Shirley L. Marks, *The JOSS Years: Reflections on an Experiment*, Santa Monica, Calif.: RAND Corporation, R-918, 1971.

[145] Tatum and Rowell's 1974 report on air base attack noted that its modeling calculations were done on an IBM System 370/158 digital computer, requiring 120 kilobytes of memory and 1.5 seconds to execute a 45-minute attack. For reference, installing the 2019 Microsoft Office applications suite on a modern computer requires 4 gigabytes of memory at a minimum.

mid-1970s onward, and RAND researchers produced numerous reports using results generated from those tools, some of which are listed below:[146]

- TALLY/TOTEM: a ground attack model that determines sortie degradation and uses those values as inputs for an air-to-air model (1973)
- AHAB: a decision analytic aid for military commanders to design air base attacks (1974)
- PROBE I: a tool to compare the relative value of assigning a fraction of sorties to different missions, such as bomber escort or air base attack (1974)
- AIDA: an air base damage assessment model for rapid examination of conventional attacks on airfields (1976)
- TSAR/TSARINA: Monte Carlo simulations that work together to analyze how attacks on air bases affect sortie generation, as well as options for improvement (1982)
- TATR: an expert system designed to aid targeteers in selecting and prioritizing targets on air bases (1983).

Of these examples, TSAR and TSARINA had the most recognition. Most of the results and major findings from TSAR and TSARINA are not publicly available, but the concept of treating the Air Base as an interactive system using computational analysis was a major RAND contribution. Countless studies benefited from the creation of TSAR and TSARINA. Donald Emerson, the model designer, alone authored a dozen reports in the 1970s and 1980s, excluding the TSAR and TSARINA user manuals.

While the technological innovation alone was impressive, it was more important for RAND to communicate its technical findings in a way that resonated with external audiences. For example, RAND's effect on Air Force decisionmaking was clearly demonstrated in 1984 when the recently retired USAFE commander General Billy M. Minter highlighted the vulnerability of USAFE's main operating bases, indirectly praising RAND's modeling and simulation research on the subject:

> It's not difficult to say that they [the airbases] are vulnerable; it's difficult to say how vulnerable. Organizations like the Rand Corporation have done a lot of research for us, and they give vulnerability estimates based on modeling and simulations [that show] we are going to suffer damage in excess of 40 percent of our support facilities. It gives me a lot of trouble.[147]

RAND also provided a modest contribution to the nascent research on remotely piloted vehicles in the 1970s. General Hap Arnold boldly stated after World War II, "We have just won a war with a lot of heroes flying around in planes. The next war may be fought by airplanes with no men in them at all."[148] Arnold was ahead of his time, but two wars later. RPVs did have a

---

[146] This list may not be exhaustive but includes some of the most notable modeling tools for ABD/A.

[147] John M. Halliday, *Tactical Dispersal of Fighter Aircraft: Risk, Uncertainty, and Policy Recommendations*, Santa Monica, Calif: RAND Corporation, N-2443-AF, 1987, p. 6.

[148] Tech. Sgt. Nadine Y. Barclay, "RPA Prophecy Fulfilled, Oldest RPA Squadron Celebrates 20 Years," Air Combat Command website, July 29, 2015.

limited role. The United States flew nearly 3,500 RPV sorties over Southeast Asia during the Vietnam War to conduct photographic, communications, and electronic reconnaissance, in addition to leaflet and chaff dropping.[149]

Although RAND's initial publications on the subject in 1971 were brief and preliminary, they included innovative analysis that paved the way for further work at RAND on the use of RPVs to attack air bases. For better or worse, Air Force interest in the subject waned over time as responsibility for RPVs shifted from SAC to Tactical Air Command (TAC). The political will was lacking, but the concept was technically viable, as demonstrated by Israel's successful use of RPVs to "fingerprint" SAM radars in the Yom Kippur War in 1973.[150]

Table 5.2 lists the major RAND contributions during this time period:

Table 5.2. Major RAND Contributions During the 1970s and 1980s

RAND Contribution	Documented in
Explored complex dynamics and trade-offs of ABD/A scenarios using nascent modeling and simulation techniques	Dadant (1973); Farquhar (1974); Tatum and Rowell (1974); Emerson (1976, 1982); Callero, Jamison, and Waterman (1983)
Assessed novel concepts for the use of RPVs	Krase (1971), Snow (1975)
Recommended dispersing resources from main operating bases and designing future aircraft to be forward deployable	Lewis, Don, Paulson, and Ware (1986); Halliday (1987)

In the next chapter, we will discuss how research on ABD/A in the 1990s and 2000s shifted after such a prolonged focus on nuclear and conventional warfare with the Soviet Union.

---

[149] Dennis Larm, *Expendable Remotely Piloted Vehicles for Strategic Offensive Airpower Roles*, Maxwell Air Force Base, Ala.: Air University Press, 1996.

[150] Larm, 1996.

6. Era of Rear Area Sanctuary for the U.S. Air Force, 1990–2009

Strategic Environment

With the end of the Cold War and the decisive American victory over Iraq in early 1991, the United States entered a period of military dominance. For roughly the next 20 years, the U.S. military enjoyed a period of rear area sanctuary, in which U.S. air supremacy ensured near complete protection for rear areas, at least from air attack. The experience during Operation Desert Storm illustrates this. There were no Iraqi air force attacks on U.S. air bases, nor were there any Iraqi special forces or Iraqi-sponsored terrorist attacks. The only strike on an airfield was a single Iraqi Scud missile impacting in an open area at Dhahran Air Base. The missile did no damage and may have been deflected there by a Patriot air defense missile.[151] American experience during Operations Deliberate Force and Allied Force further validated this sanguine view of the threat. The truck bombing of the USAF barracks at Khobar Towers in Saudi Arabia in 1996 was not on an air base, but it did offer a cautionary note about the vulnerability of American facilities to terrorist attack—still, most viewed the attack's relevance to air base defense limited to proper entry control and perimeter security.

Thus, the consensus view in the defense community during most of this period was that the threat to air bases would be minimal in conflicts with regional powers such as Iraq and Serbia, and that, in the unlikely event of a conflict in Korea, existing defensive measures were sufficient. This changed somewhat during Operations Enduring Freedom and Iraqi Freedom, when rocket and mortar attacks on air bases became quite common. There also were attacks on air base entry control points during this period. Although these threats presented significant challenges for USAF Security Forces and resulted in some important innovations, such as the Army's Counter–Rocket, Artillery, and Mortar (C-RAM) system, they did not fundamentally alter the perception in the USAF and broader defense community that U.S. rear areas (including air bases) were largely safe from adversary attack. It was only at the very end of this period, during 2008 and 2009, that analysts, planners, and some senior military leaders began to express alarm about the potential vulnerability of air bases. More will be said on these attitudes and events below.

Perhaps the most critical change in the strategic environment occurred in the thinking of potential adversaries who were dumbstruck by the U.S. air dominance during Operation Desert Storm and began to consider means to counter the U.S. ability to project power. The most consequential of these reactions emanated from the Chinese military.

---

[151] Bernard Rostker, *Information Paper: Iraq's SCUD Ballistic Missiles*, Washington, D.C.: U.S. Department of Defense, interim paper, July 25, 2000.

Chinese military leaders and planners thought deeply and creatively about the lessons of the American victory over Iraq in 1991, particularly the contribution of U.S. airpower. Chinese military theorists understood that China was highly vulnerable to a similar air campaign.[152] In response, during the 1990s the PLA began to develop concepts and weapon systems designed to counter American air power, including advanced integrated air defenses; expanded intelligence, surveillance, and reconnaissance (ISR); and highly accurate conventional ballistic and cruise missiles. The missile systems and associated concepts of operation were conceived as asymmetric means to disrupt the U.S. advantage.[153] American analysts first noticed these capabilities in the late 1990s,[154] and multiple publications over the next decade offered greater detail on Chinese thinking and capabilities.[155]

We will say more about the China threat toward the end of this chapter, but it was not a concern in 1990. Rather, the primary focus of U.S. defense planning at the beginning of this period was ensuring that American forces could prevail in two near-simultaneous regional conflicts while conducting various peace enforcement operations in the Middle East and Balkans, all while undergoing a substantial post–Cold War reduction in forces, bases, and funding.[156]

In this chapter, we first discuss the period of peace operations that lasted from 1990 to 2001. Although the events of 9/11 did not increase conventional threats to USAF bases, the strategic environment and policy priorities so changed with the war on terror that it is most appropriate to treat the period from 2001 to 2009 separately. Also, the very end of this period (2008–2009) saw

[152] For more on Chinese reactions to the first Gulf War, see Roger Cliff, Mark Burles, Michael S. Chase, Derek Eaton, and Kevin L. Pollpeter, *Entering the Dragon's Lair: Chinese Antiaccess Strategies and Their Implications for the United States*, Santa Monica, Calif.: RAND Corporation, MG-524-AF, 2007, especially pp. 28–32.

[153] For an analysis arguing that Chinese advances are disruptive innovations, see Vick, 2015, pp. 32–38.

[154] The first two public studies to address this threat were Mark A. Stokes, *China's Strategic Modernization: Implications for the United States*, Carlisle, Pa.: Strategic Studies Institute, U.S. Army War College, September 1999; and John Stillion and David T. Orletsky, *Airbase Vulnerability to Conventional Cruise-Missile and Ballistic-Missile Attacks: Technology, Scenarios, and U.S. Air Force Responses*, Santa Monica, Calif.: RAND Corporation, MR-1028-AF, 1999.

[155] See David A. Shlapak, David T. Orletsky, and Barry Wilson, *Dire Strait? Military Aspects of the China-Taiwan Confrontation and Options for U.S. Policy*, Santa Monica, Calif.: RAND Corporation, MR-1217-SRF, 2000; Thomas Christensen, "Posing Problems Without Catching Up: China's Rise and Challenges for U.S. Security Policy," *International Security*, Vol. 25, No. 4, Spring 2001; and Christopher J. Bowie, *The Anti-Access Threat and Theater Air Bases*, Washington, D.C.: Center for Strategic and Budgetary Assessments, 2002.

[156] Although published somewhat later, RAND's *Strategic Appraisal 1997* captures the defense planning environment of the 1990s in which peace operations, debates over threat versus capabilities-based planning, budget pressures, infrastructure reductions and regional threats all vied for the attention of policymakers. See Zalmay M. Khalilzad and David Ochmanek, eds., *Strategic Appraisal 1997: Strategy and Defense Planning for the 21st Century*, Santa Monica, Calif.: RAND Corporation, MR-826-AF, 1997. For an influential analysis of the two regional conflict planning requirement, see Christopher Bowie, Fred Frostic, Kevin Lewis, John Lund, David Ochmanek, and Phillip Propper, *The New Calculus: Analyzing Airpower's Changing Role in Joint Theater Campaigns*, Santa Monica, Calif.: RAND Corporation, MR-149-AF, 1993.

yet another shift in strategic thinking as U.S. planners began to understand the significance of Chinese military modernization for the American way of war.

1990–2000

The Cold War is generally considered to have ended when the Soviet Union formally dissolved on December 26, 1991, but, as a practical matter, the Soviet/Warsaw Pact threat to NATO had greatly eroded by November 1989, when the Berlin Wall fell. By late 1989, in recognition of the reduced threat to the United States, General John Chain, Commander, Strategic Air Command, recommended that SAC end its continuous airborne command post mission. Initially reluctant to do so, defense leaders ultimately approved this recommendation. On July 24, 1990, SAC flew its last command post mission, ending a remarkable 30 years during which a SAC EC-135 command post aircraft was airborne somewhere over North America 24 hours a day, seven days a week.[157] Therefore, 1990 is a better date to designate the end of Cold War class threats to USAF bases and the beginning of the next strategic era.[158]

The central strategic problem facing the United States in this era was (ironically) caused by the dissolution of the Soviet Union: the potential for instability in its former republics and the risk that nuclear weapons in Kazakhstan and other former republics might be given or sold to terrorists. A secondary concern was that regional powers such as Iraq, Iran, or North Korea might feel more free to use military power than they did during the Cold War.

This latter scenario came to pass in August 1990, when Iraq invaded Kuwait. The policy challenges during this period were far from easy, but American leaders could make military choices without worrying about Soviet reactions or escalatory risks. As the sole superpower, the United States took a leadership role, diplomatically and militarily, that was only partially constrained by Soviet/Russian attitudes and preferences.

Operations Desert Shield and Desert Storm established a template for a new American way of war that would be followed in subsequent interventions over the next two decades, beginning with the rapid deployment of air, naval, and light ground elements to Saudi Arabia.[159] This was followed by a massive multi-month deployment of joint forces to the Persian Gulf area. Although American defense planners respected Iraqi air defenses and ground forces, they did not believe that Iraqi aircraft, missiles, or special operations forces (SOF) could threaten USAF bases. American air superiority was taken for granted, and the bases operated in sanctuary from attack. As a result, large numbers of aircraft were deployed to forward bases, where they

---

[157] Ground alert for bombers and tankers ended one year later, in September 1991. See Strategic Air Command, Office of the Historian, 1991, pp. iii and 46–47.

[158] Melvyn Leffler, a prominent Cold War historian, also designates 1990 as the last year of the Cold War. See Melvyn P. Leffler, *For the Soul of Mankind: The United States, the Soviet Union, and the Cold War*, New York: Hill and Wang, 2007.

[159] For more on this new American way of war, see Vick, 2015, pp. 11–18.

typically were parked close together on crowded ramps. Only fighter aircraft at a minority of bases had revetments or hardened shelters for protection. American military planners proved to be correct. No Iraqi aircraft or SOF attacks were launched against allied air bases. One SCUD missile impacted at Dhahran Air Base, creating a crater in an open area but otherwise causing no damage.[160] The only SCUD attack of any consequence was the strike on a U.S. barracks in Al Khobar that killed 27 and wounded 98. Given the inaccuracy of Iraqi SCUDs, the strike was pure luck but nonetheless tragic.[161]

U.S. air superiority was so complete in the 1990s that the USAF was able to operate in a similar manner in no-fly zone and power projection operations in both the Balkans and Persian Gulf. In Operations Northern and Southern Watch, Deny Flight, Deliberate Force, and Allied Force, U.S. and partner-nation air forces operated from forward bases with minimal risk of adversary strikes against these bases.[162]

For roughly 20 years after the Soviet Union dissolved in 1991, the "battle of the airfields" was a one-sided affair for the United States. Enjoying overwhelming air superiority, the only issue for the U.S. military was when and how to attack adversary airfields during conventional conflicts. Thus, during Operations Desert Storm, Allied Force, Enduring Freedom, Iraqi Freedom, and Odyssey Dawn, adversary airfields were among the earliest targets struck by U.S and coalition forces. The USAF and sister service aviation elements had little to fear from enemy attack on airfields or aircraft carriers. Terrorist and other ground threats to air bases were taken seriously, but perimeter defenses, strong entry control points, patrols, and host-nation security forces were more than adequate to handle these threats during the conventional phase of the conflicts listed above.[163]

2001–2010

Following the September 11, 2001, attacks on the United States, the central strategic problem facing the United States was the defeat of al Qaeda and its Taliban supporters. American forces moved rapidly to the Middle East and South Asia littoral, beginning offensive air operations in Afghanistan on October 7. Facing no threat from the Afghan air force, terrorist attacks against U.S. bases in the region were the only concern for planners. Once the Taliban regime fell and

---

[160] The air base may have been the intended target, or the missile may have been deflected there by an intercepting Patriot missile. See Rostker, 2000.

[161] R. W. Apple, Jr., "War in the Gulf: Scud Attack; Scud Missile Hits a U.S. Barracks, Killing 27," *New York Times*, February 26, 1991.

[162] Operation Odyssey Dawn, although occurring in 2011 during the post-9/11 era, also followed this model where USAF and partner air forces conducted operations against Libya from sanctuary bases in southern Europe. See Karl P. Mueller, ed., *Precision and Purpose: Airpower in the Libyan Civil War*, Santa Monica, Calif.: RAND Corporation, RR-676-AF, 2015.

[163] The 1996 truck bombing of Khobar Towers in Saudi Arabia illustrated gaps in USAF and Saudi security, but because the towers were in the city of Khobar and not on an air base, this incident is not evidence of air base security failures.

Kabul was in allied hands, the nature of the conflict shifted to counterinsurgency and counterterrorism operations. These required the USAF to operate from Bagram and other airfields in Afghanistan. This was a significant shift from operations conducted from bases in well-policed and relatively secure nations such as Qatar. Air base defenses had to prepare for rocket, mortar, and commando-style attacks and suicide bombings against entry points. Although they never achieved the scale or effectiveness of Viet Cong attacks on USAF bases during the Vietnam War, Taliban insurgents regularly conducted small rocket attacks on coalition air bases in Afghanistan between 2001 and 2014. The attacks were more of a nuisance that serious threat. The one exception was the September 2012 attack on Camp Bastion, in which insurgents penetrated the perimeter and destroyed six U.S. Marine Corps A-8 Harrier fighter aircraft with grenades and rocket-propelled grenades.[164]

The invasion of Iraq in 2003 (Operation Iraqi Freedom) was similar to Desert Storm. In each case, the United States was able to deploy forces into theater without disruption by enemy offensive action and initiate combat at a time and place of its own choosing.[165] U.S. air bases faced no threat from enemy action. That changed after Baghdad fell in April 2003 and the United States transitioned to occupation duties and then to counterinsurgency operations. These required the USAF to operate from air bases inside Iraq (e.g., Joint Base Balad), which became targets for frequent insurgent rocket or mortar attacks. In response, the United States and its coalition partners devoted significant resources to defending airbases: enhancing perimeter fencing and sensors, building fighting positions and fortified entry points, using Hesco barriers to create protective walls, creating personnel shelters to use during rocket attacks, patrolling the standoff weapon footprint, and using airborne platforms for ISR. These efforts were largely successful in minimizing casualties and equipment damage. That said, between 2003 and late 2006, insurgents still managed to launch more than 1,500 standoff attacks against USAF bases.[166]

Operation Odyssey Dawn, the 2011 U.S. and NATO air operation against Libya, was more in line with the post–Cold War model, with the launching of air attacks from sanctuary bases in southern Europe against Gaddafi's regime. Operation Inherent Resolve, the operation against the Islamic State of Iraq and Syria (ISIS; 2014–today), is a hybrid with air operations conducted from sanctuary bases against ISIS elements conducting both irregular and conventional combat operations.

---

[164] Alissa J. Rubin, "Audacious Raid on NATO Base Shows Taliban's Reach," *New York Times*, September 16, 2012, online.

[165] For more on the Operation Iraqi Freedom air war, see Benjamin S. Lambeth, *The Unseen War: Allied Air Power and the Takedown of Saddam Hussein*, Annapolis, Md.: Naval Institute Press, 2013.

[166] Paul M. Thobo-Carlsen, "A Canadian Perspective on Air Base Ground Defense: Ad Hoc Is Not Good Enough," in Shannon W. Caudill, ed., *Defending Air Bases in an Age of Insurgency*, Maxwell Air Force Base, Ala.: Air University Press, 2014, p. 53.

Overview of RAND Research on Air Base Defense and Attack: 1990–2009

As noted in earlier chapters, RAND's defense research agenda reflects the policy priorities of DoD and other government sponsors. With the end of the Cold War, there was little DoD concern about air base resiliency but considerable interest in larger force planning problems, peace operations, and understanding the new security environment.

Thus, it comes as no surprise that the number of RAND reports on ABD/A dropped considerably after 1990, with just 18 reports published on ABD/A between 1990 and 2009, a dramatic drop from the 49 reports published during the 1980s. A closer look at the 1990s reports suggests the drop is even greater than it initially appears. Six reports were published in 1990. Given the normal delays associated with reviewing and publishing research, the work was likely conducted prior to the end of the Cold War. For example, two of the 1990 reports explored the most effective means to attack Soviet and Warsaw Pact bases, a topic rendered moot by the end of the Cold War. The other four 1990 reports were updates for the TSAR/TSARINA models, which continued to be used some in the early 1990s. In 1991, RAND published a report on munitions options for attacking Warsaw Pact bases, and in 1992, RAND published three reports on NATO air base defense. There were no ABD/A reports published in 1993 or 1994. RAND published three ABD/A reports in 1995. One was a lessons learned effort that assessed the effectiveness of Iraqi hardened aircraft shelters (some of best in the world) against U.S. precision weapons. The other two returned to the topic of ground threats to air bases, first addressed by RAND in 1967 in the context of Vietnam. In 1999, in the context of a project investigating adversary asymmetric strategies, a RAND report explored the possibility that future adversaries would use GPS-guided missiles to attack U.S. aircraft parked in the open. Finally, in 2000, in a study sponsored by the Smith Richardson Foundation (as opposed to DoD), RAND returned to campaign-level analyses and air base attack modeling with its first quantitative analysis of a China-Taiwan war.

RAND published only three reports on air base defense between 2001 and 2009 and no reports on air base attack. In 2005, a team led by William Stanley produced a documented briefing that presented findings from an analysis of conventional ballistic and cruise missile threats to overseas bases.[167] In 2009, RAND updated and expanded the earlier analysis of a China-Taiwan conflict. The third report (also produced in 2009), by RAND analysts Roger Cliff, Aidan Kirby Winn, and David Ochmanek, summarized insights from the Pacific Vision wargame.

Finally, although only briefly touching on air base attacks (and therefore not in our dataset), the Roger Cliff et al. report *Entering the Dragon's Lair* helped American audiences better understand Chinese thinking about offensive air and missiles campaigns and, in particular, the importance of preemptively attacking enemy air bases. We do not include this report in our

---

[167] Other authors were Robert Uy, Carl Rhodes, Rich Mesic, Dan Norton, and John Tonkinson.

dataset, but it should be recognized for helping build a consensus regarding the PLA threat to USAF bases.[168]

The sparse publications between 2001 and 2009 reflected the shift of RAND's analytical focus to the problems of counterterrorism and counterinsurgency and the prioritization of research most directly helpful to ongoing combat operations in Iraq and Afghanistan. The publications record is, however, somewhat misleading regarding air base defense research during those years. In 2008, RAND Project AIR FORCE (PAF) was deeply involved in two projects that would—through wargames and briefings—contribute to a heightened awareness among senior USAF and DoD leaders regarding emerging threats to forward air bases.

The first of the 2008 projects was sponsored by the Commander, 13th Air Force at Hickam AFB, Hawaii. The project, "Implications of Adversary Military Modernization for USAF Basing and Operations," focused on the evolving missile threat to USAF bases in the Pacific.[169] Team member John Stillion updated and expanded the 1999 Stillion and Orletsky analysis to better reflect advances in guidance systems and submunitions, using a simple model to assess the effects of small submunitions on aircraft parked in the open at Pacific Air Forces (PACAF) bases. This analysis was widely briefed to USAF and DoD audiences.

The second project was a game that supported PACAF commander General Carrol "Howie" Chandler's Pacific Vision initiative, a series of games and analyses that he sponsored to better understand emerging challenges in the Pacific region. Chandler reached out to Andrew Hoehn (at the time RAND's Vice President for PAF) and Thomas Ehrhard at the Center for Strategic and Budgetary Analysis to build a game that would explore future threat scenarios. Ehrhard and Robert Martinage had developed a gaming structure and portfolio analysis tool that proved quite successful in previous CSBA games. Pacific Vision used this game structure. RAND provided the substantive input, combining analysis from Stillion and others with a conflict scenario written by Asia specialist Roger Cliff.

The Pacific Vision game was conducted at Hickam Air Force Base over several days in August 2008 and included participants from all branches of the U.S. military, as well as Royal Australian Air Force and Naval personnel. The Red Team included weapon systems experts from the National Air and Space Intelligence Center, analysts from the Office of Naval Intelligence, USAF officers, and Asia regional specialists from RAND. Game adjudication was done by a RAND team headed by Ochmanek, with RAND senior engineer Jeff Hagen acting as the lead combat adjudicator. The game benefited from the active participation of three distinguished senior mentors: Admiral Dennis Blair (U.S. Navy, retired), former commander of U.S. Pacific Command; Lt Gen Charles Heflebower (USAF, retired) and Andrew Hoehn from RAND.

---

[168] Cliff et al., 2007, pp. 62–64.

[169] Project team members were Lauren Caston, Roger Cliff, Michael Lostumbo, David Shlapak, John Stillion, and Alan Vick (principal investigator).

The game was highly influential in two ways. First, Pacific Vision participants overwhelmingly agreed that the game highlighted challenges that had not been adequately appreciated. The game and supporting analyses led General Chandler to pursue a variety of mitigation efforts to address the issues raised. Second, two key game leaders (Ehrhard and Ochmanek) entered government service not long after the game. Ehrhard was the first, joining the Senior Executive Service in late 2008 as special assistant to General Norton Schwartz, the USAF Chief of Staff. In that position, Ehrhard was able to share insights from Pacific Vision within the Pentagon and advocate for new concepts and programs. David Ochmanek also went into government, serving as Deputy Assistant Secretary of Defense, Force Development, from 2009 to 2014. In this position, he sponsored multiple wargames along the lines of Pacific Vision, led at least one task force on air base vulnerabilities and, more broadly, was an active voice for improvements. For example, in a 2014 interview Ochmanek observed that "Planners worry about what happens to our forward-based forces when they're inside the threat range from ballistic missiles and cruise missiles if those weapons are accurate and if they're deliverable in large numbers." Ochmanek also noted that gaming and analysis has shown "promising results" from "dispersing the force more radically on and across airfields."[170]

Table 6.1 provides an overview of the policy and analytical objectives of these reports. Fifteen addressed some aspect of air base defense, and four continued the body of RAND analysis devoted to air base attack.

The 1995 report on Iraqi aircraft shelters would be RAND's last publication on the topic of attacking enemy air bases for 20 years.[171] There were, however, at least two PAF analyses of air base attack conducted roughly 10 to 15 years later that were shared in briefings but never published. In the 2005–2010 timeframe, RAND engineer Donald Stevens assessed whether USAF and U.S. Navy conventional air and missile strikes could significantly reduce sortie generation in a conflict with a near-peer adversary. In a FY 2009 study for the USAF Quadrennial Defense Review Office, analysts David Frelinger and Barry Wilson used the Joint Integrated Contingency Model (JICM) to assess the relative effectiveness of three concepts of operation (and associated program options) for the USAF in a major conflict with a near peer.[172]

---

[170] Quoted in Marcus Weisgerber, "Pentagon Debates Policy to Strengthen, Disperse Bases," *Defense News*, April 13, 2014. Although Ochmanek did not mention it by name, he was likely referring to the Office of the Secretary of Defense–sponsored RAND study on cluster basing published the same year as this interview. Chapter 7 will briefly discuss this study.

[171] Eric Heginbotham, Michael Nixon, Forrest E. Morgan, Jacob L. Heim, Jeff Hagen, Sheng Li, Jeffrey Engstrom, Martin C. Libicki, Paul DeLuca, David A. Shlapak, David R. Frelinger, Burgess Laird, Kyle Brady, and Lyle J. Morris, *The U.S.-China Scorecard: Force, Geography, and the Evolving Balance of Power, 1996–2017*, Santa Monica, Calif.: RAND Corporation, RR-392-AF, 2015.

[172] The project "Supporting USAF Participation in the QDR" was sponsored by then–Major General James Hunt (USAF/CVAQ) and then–Brigadier General Mark Ramsay (USAF/A8X). Project officers were Carl Rehberg (USAF/CVAQ) and Max Hanessian (USAF/A8XP). James Brooks, associate director of the Quadrennial Review Office (USAF/CVAQ), also played a major role in project oversight. The project assessed USAF program requirements for both major conflicts and irregular warfare. The RAND project team was made up of Mike Boito,

They modeled USAF standoff attacks on enemy air bases (using a large number of air-launched cruise missiles). Given the cyclical nature of research in this area, one cannot rule out future work on offensive options.

Table 6.1. Top Policy/Analytical Objectives of RAND Reports on Air Base Defense and Attack: 1990–2009

Policy/Analytical Objective	Number of RAND Reports
Identify most effective means to defend airfields	8
Conduct air operations during and after conventional and chemical attack	4
Identify most effective means to attack enemy airfields	3
Protect air bases from ground attack	2
Identify most effective means to attack and defend airfields	1
Identify most effective means to attack enemy hardened aircraft shelters	1

A Sampling of Reports

In this section, we will discuss four RAND reports from this period that reflect the breadth of research on ABD/A:

- *"Check Six Begins on the Ground": Responding to the Evolving Ground Threat to U.S. Air Force Bases* (1995)
- *Airbase Vulnerability to Conventional Cruise-Missile and Ballistic Missile Attacks: Technology, Scenarios, and U.S. Air Force Responses* (1999)
- *Dire Strait? Military Aspects of the China-Taiwan Confrontation and Options for U.S. Policy* (2000)
- *A Question of Balance: Political Context and Military Aspects of the China-Taiwan Dispute* (2009).

The first report, *Check Six Begins on the Ground*, was part of a larger PAF study investigating asymmetric strategies that future adversaries might use to counter U.S. dominance in the air. Authors David Shlapak and Alan Vick asked whether a future adversary might use special operations forces, terrorist cells, or insurgents to neutralize or at least blunt U.S. air power. "Taking advantage of readily available forces and technologies, [the adversary] could hope to reduce the effectiveness of U.S. air operations, at least temporarily, by destroying high-value assets or disrupting sortie generation."[173] The authors conclude that many potential adversaries have the force elements necessary to conduct ground attacks on air bases and that many of the key technologies (e.g., sniper rifles, man-portable air-defense systems

---

David Frelinger, Adam Grissom, Jessica Hart, Angel Martinez, Heather Peterson, Fred Timson, Alan Vick (PI), and Barry Wilson.

[173] David A. Shlapak and Alan Vick, *"Check Six Begins on the Ground": Responding to the Evolving Ground Threat to U.S. Air Force Bases*, Santa Monica, Calif.: RAND Corporation, MR-606-AF, 1995, p. xiii.

[MANPADS], accurate mortars, GPS receivers, night vision devices) were widely available in the mid-1990s. They emphasize that the greatest threat is from standoff attacks using either traditional weapons, such as mortars and rockets, or newer technologies, such as anti-tank guided missiles (ATGMs), all of which can be lethal against aircraft parked in the open. Most worrisome is the U.S. reliance on small numbers of high-value aircraft, such as Airborne Warning and Control System (AWACS), Joint Surveillance Target Attack Radar System (JSTARS), and bombers.

Finally, the authors note that the threat is not just theoretical, but one demonstrated in more than 600 ground attacks on airfields in ten separate conflicts occurring between 1940 and 1992. A companion volume, *Snakes in the Eagle's Nest: A History of Ground Attacks on Air Bases*, analyzes these attacks in detail.[174] The authors of *Check Six Begins on the Ground* divide their recommendations into two categories. Against the penetrating threat, they recommend (1) improvements in on-base surveillance and on approaches to bases, (2) procurement of up-armored HMMWVs for quick reaction forces and patrols, and (3) weapon mounts for security force vehicles so that more can be equipped with grenade launchers and machine guns. Against the standoff threat, they envision a three-pronged strategy: (1) Confound adversary mission planning through deception, decoys, camouflage and rotation of aircraft through multiple bases, (2) detect and defeat the adversary before they can launch attacks, and (3) protect key assets through hardening, including the development of hardened shelters for large, high-value aircraft.

The second report, *Airbase Vulnerability to Conventional Cruise-Missile and Ballistic Missile Attacks: Technology, Scenarios, and U.S. Air Force Responses*, was published in 1999. This report was written under the auspices of a larger project exploring the role of air and space power in future conflicts. The authors sought to understand whether the "current USAF operational concept of high-tempo, parallel strikes from in-theater bases could be put in jeopardy" by the proliferation of GPS-guided missiles and submunition technology.[175] The authors concluded that in a future Iraq scenario in which the USAF again based aircraft in the open at Dhahran, Doha, Riyadh Military, and Al Kharj air bases, an attack using 30 GPS-guided ballistic and 30 cruise missiles armed with small (roughly grenade size) submunitions could achieve a 0.9 Pk (probability of kill) against aircraft in the open. Although they use the specifics of basing during Operation Desert Storm (e.g., aircraft parking density, size of parking ramps), they argue that this threat to USAF operations could be executed in any theater, particularly during expeditionary operations to austere bases. To address this potential vulnerability the authors suggest USAF responses in three areas: (1) passive defenses (including hardened aircraft shelters), (2) simple near-term counter cruise missile defenses, and (3) dispersal to highway landing strips. Looking to the future, the report also considers the possibility of operating from

---

[174] Vick, 1995.

[175] Stillion and Orletsky, 1999, p. iii.

greater range, proposing a medium bomber with a cruise speed of 1,000 knots, weight of 290,000 to 350,000 pounds, payload of 15,000 to 20,000 pounds, and unrefueled range of 3,250 nm.

The third and fourth reports, *Dire Strait? Military Aspects of the China-Taiwan Confrontation and Options for U.S. Policy (2000)* and *A Question of Balance: Political Context and Military Aspects of the China-Taiwan Dispute* (2009), both present campaign-level analyses of a China-Taiwan conflict.

Dire Strait? uses qualitative and quantitative methods to explore a conflict set in 2005. The authors used the JICM to simulate the war.[176] The report presents seven major findings related to the air and naval battles. The first of the findings notes that "Taiwan's air bases must remain operable so that the ROCAF's [Republic of China Air Force's] fighter force can keep up the fight against the superior numbers of the PLA Air Force (PLAAF)." The report identifies four vulnerabilities that were highlighted in their analysis: (1) above-ground fuel storage tanks and parking areas for fuel trucks, (2) unhardened engine and avionics repair shops, (3) operating surfaces (runways and taxiways), and (4) the potential use of chemical agents against air bases.[177] In response to these vulnerabilities, the authors "recommend increased attention to passive defense and rapid-reconstitution measures."[178]

A Question of Balance considers a conflict in the 2010–2015 timeframe. The authors find that the military advantage had shifted dangerously to Beijing because of the rapid advances its air and missile forces had made in the intervening years. The report summary regarding Chinese missile force improvements is worth reproducing in full:

> We assessed the potential impacts of these weapons [short-range ballistic missiles] against Taiwan's air bases. Using Monte Carlo techniques to model these attacks, we found that, depending on missile accuracy, between 90 and 240 SRBMs [short-range ballistic missiles]—a number well within the range of estimates of the number of launchers China will field in the near future—could, with proper warheads, cut every runway at Taiwan's half-dozen main fighter bases and destroy essentially all of the aircraft parked on ramps in the open at those installations. By so doing, China could knock the Republic of China Air Force (ROCAF) out of the war for long enough to launch large-scale air raids on Taiwan intended to destroy any aircraft parked in shelters, as well as other hardened targets. Success in this gambit would suppress ROCAF operations indefinitely and lay Taiwan open to further Chinese air attacks.[179]

The report addresses other aspects of the Taiwan-China military balance, but, for our purposes, two conclusions are most salient. First, the authors note that redesigning "Taiwan's air

---

[176] For an overview of JICM, see Shlapak, Orletsky, and Wilson, 2000, pp. 63–84.

[177] Shlapak, Orletsky, and Wilson, 2000, p. 33.

[178] Shlapak, Orletsky, and Wilson, p. xvi.

[179] David A. Shlapak, David T. Orletsky, Toy I. Reid, Murray Scot Tanner, and Barry Wilson, *A Question of Balance: Political Context and Military Aspects of the China-Taiwan Dispute*, Santa Monica, Calif.: RAND Corporation, MG-888-SRF, 2009, p. xv.

defenses to 'ride out' heavy strikes on its bases and other installations can complicate Chinese planning and reduce the leverage that Beijing can derive from its offensive forces."[180] Second, regarding the potential for missile strikes on U.S. bases in East Asia, the authors observe that clearly communicating "to Beijing the consequences of attacking U.S. bases and forces in East Asia in terms of counterstrikes on the Chinese mainland has the potential to enhance deterrence."[181]

In summary, although 1990–2009 saw a significant shift in the geopolitical environment, U.S. policy priorities and the emphasis of RAND defense policy research, RAND never entirely abandoned the field of ABD/A research. Indeed, original and enduring contributions were made in several areas which we discuss in the next section.

RAND Contributions

The collapse of the Soviet Union and end of its dominance over Eastern Europe, combined with the stunning American coalition victory over Iraq in 1991, created a degree of triumphalism among Americans.[182] Some airmen and air power analysts saw Operation Desert Storm as heralding a new era in which air power would be the dominant arm of the American military and would reign unchallenged by adversary air forces for many years.[183] Although this view was appealing to many or perhaps most airmen, the institutional Air Force was more cautious. During the 1990s, senior USAF leaders sought to understand how future adversaries might adapt to these American advantages, and the USAF sponsored multiple RAND studies on emerging threats in the early to mid-1990s, including two on ground threats to air bases and one on the threat that GPS-guided missiles posed to air bases.

Table 6.2 highlights five RAND reports completed between 1990 and 2009 that made enduring contributions to the ABD/A studies and related defense policy. The first three were all sponsored by the USAF; the two Shlapak et al. studies on a Taiwan-China conflict (2000 and 2009) were sponsored by the Smith Richardson Foundation. These five reports were all summarized in the previous section, so we will just add a few additional observations here.

The Shlapak and Vick reports on ground threats to air bases were notable in two regards. The few prior studies on the topic of ground threats to air bases (e.g., those conducted during the Vietnam War) viewed the issue purely as a near-term tactical problem to be solved in the context

---

[180] Shlapak et al., 2009, p. xix.

[181] Shlapak et al., 2009, p. xix.

[182] Perhaps the most famous example is Francis Fukuyama, *The End of History and the Last Man*, New York: Free Press, 1992.

[183] Lt Gen David Deptula, RAND analyst Benjamin Lambeth, and USAF historian Richard Hallion are the best known proponents of this view. For more on their views and other airpower narratives, see Alan J. Vick, *Proclaiming Airpower: Air Force Narratives and American Public Opinion from 1917 to 2014*, Santa Monica, Calif.: RAND Corporation, RR-1044-AF, 2015, pp. 67–75.

of a single conflict. *Check Six Begins on the Ground* was the first analytical effort to consider ground threats in the context of adversary strategy. First, the report explained how ground attacks on air bases could serve strategic purposes, acting as an asymmetric counter to American air power and possibly creating "strategic events" that might undermine American public support for military intervention. Second, the report illustrated how diverse adversaries (from terrorists to major powers' special operations forces) might use current and emerging technologies to successfully attack air bases in a variety of conflict scenarios. Finally, the historical chapter in *Check Six* (and its companion volume, *Snakes in the Eagle's Nest*) made a strong empirical case that the standoff threat to air bases (e.g., mortars and rockets) was much greater than the penetrating threat that had previously been emphasized in USAF Security Force doctrine. USAF Security Force leaders were sufficiently convinced by this analysis that USAF Force Protection doctrine (AFDD 2-4.1) was changed in 1996 to emphasize the standoff threat and both RAND reports were cited in that and subsequent revisions. The authors, to their great astonishment, also enjoyed the rare distinction of being quoted by name in multiple epigraphs in AFDD 2-4.1, along with such notables as Winston Churchill, Frederick the Great, and USAF Chief of Staff General Ronald R. Fogleman.[184] These two reports became part of the USAF Security Force canon, widely used in courses and cited dozens of times in the most comprehensive USAF treatment of air base defense from ground attack.[185]

The 1999 Stillion and Orletsky report became a standard reference on the topic of air base attack for several reasons. Similar to the Shlapak and Vick reports on ground threats (Shlapak and Vick, 1995; Vick; 1995), Stillion and Orletsky's report began with a recognition of the large role that air power played in the 1991 victory over Iraq and a desire to ensure that USAF operating practices would remain robust in the face of enemy countermeasures. Stillion, who served as a USAF RF-4 weapon system officer during the Cold War, was struck by the contrast between USAF bases in Europe (characterized by hardening, dispersion, and camouflage) and USAF operating locations during Operation Desert Storm (characterized by aircraft densely parked in the open with no protection). The latter basing scheme proved to be viable in the face of Iraq's small and inaccurate missile force, but the authors wondered whether advances in missile guidance systems, especially GPS, might give future adversaries a means to counter U.S. air power without the cost and difficulty of fielding a world-class air force. Their analysis demonstrated that projected accuracy improvements would make submunition-equipped cruise and ballistic missiles deadly threats to aircraft parked in the open, making expeditionary operations following the Operation Desert Storm template highly risky. The report seamlessly integrated strategy, threats, and technologies into an engaging narrative. More importantly, their

---

[184] U.S. Air Force, *Force Protection*, Air Force Doctrine Document 2-4.1, Maxwell Air Force Base, Ala. Air Force Doctrine Center, October 29, 1999.

[185] See Shannon Caudill, ed., *Defending Air Bases in An Age of Insurgency*, Maxwell Air Force Base, Ala.: Air University Press, 2014; and Shannon Caudill, ed., *Defending Air Bases in An Age of Insurgency: Volume II*, Maxwell Air Force Base, Ala.: Air University Press, 2019.

operational analysis was simultaneously rigorous and elegantly simple. There was nothing else like it in the air base defense literature, so the 1999 Stillion and Orletsky report became a, if not the, standard reference on the topic for the next two decades.

The two reports that Shlapak and team produced on a Taiwan-China conflict—*Dire Strait?* (2000) and *A Question of Balance* (2009)—were naturally of great interest to Asia specialists, addressing a range of political, strategic, and operational issues confronting Taiwan and the United States. Particularly noteworthy is the change in outlook between the 2000 and 2009 reports, with the latter report expressing much greater pessimism about the outcome of a future conflict because of China's rapid advances in air and missile capabilities. These reports were also influential in the broader defense planning community, offering insights regarding how the PLA might use ballistic missiles to attack USAF bases in East Asia, how long air bases might be closed by these attacks and, of greatest consequence, how air base attacks would affect the larger air campaign. *Dire Strait?* and *A Question of Balance* both were unique in their integration of air base missile attacks into a campaign-level model, JICM. Although higher-resolution air base attack models were developed in prior years (e.g., TSAR and TSARINA), those simulations were used to generate simple metrics, such as sorties generated, but they did not integrate those metrics into a campaign-level model. Similarly, other campaign-level models have existed prior to and after JICM, but those models typically treated air base attacks in highly stylized ways that offered few, if any, insights on the relative effectiveness of specific attack vectors on air operations. JICM offered a happy medium between these two extremes. To the best of our knowledge, Shlapak and his co-authors were the first to fully incorporate air base attacks with modern ballistic missiles into a campaign-level analysis. Finally, *A Question of Balance* offered the most detailed open source modeling of air base attack yet published.

Table 6.2. RAND Contributions to Air Base Defense and Attack Research, 1990–2009

RAND Contribution	Documented in
Analysis of air base ground attack as an adversary asymmetric strategy	Shlapak and Vick (1995)
Comprehensive history of ground attacks on air bases	Vick (1995)
Detailed analysis of GPS-guided missile threat to USAF bases	Stillion and Orletsky (1999)
Integration of missile attacks on air bases in campaign-level model	Shlapak et al. (2000, 2009)

In the next chapter, we describe the return of major power conflict as a planning priority, the rise of A2/AD capabilities, and return of air base survivability as a major emphasis of RAND research.

7. Anti-Access Threat to U.S. Bases Reinvigorates Analysis of Air Base Defense, 2010–2020

Strategic Environment

Although the United States remains involved in combat operations against violent extremists in the Middle East, South Asia, and North Africa, there is growing recognition that the principal strategic problem facing the nation is the emergence of two major power adversaries. The 2018 National Defense Strategy states:

> Long-term strategic competitions with China and Russia are the principal priorities for the Department, and require both increased and sustained investment, because of the magnitude of the threats they pose to U.S. security and prosperity today, and the potential for those threats to increase in the future.[186]

Although forward-looking analysts, senior military officers, and DoD officials recognized the emerging Chinese threat in publications, wargames, and studies as early as 1999 and increasingly between 2008 and 2017, neither Congress nor the executive branch made it a planning priority, nor was there a public consensus in support of increased defense spending to counter peer-level threats. Increasingly aggressive Chinese actions in the South China Sea and Russian seizure of the Crimea and invasion of eastern Ukraine in 2014 have changed elite and public attitudes toward these threats, and there now is a growing consensus that China and Russia represent significant threats to U.S. interests.

The essence of the emerging Chinese threat to USAF bases is found in a strategy that emphasizes the importance of surprise and preemption and in a force structure that includes the largest and most advanced conventional ballistic missile force in the world, as well as ground- and air-launched long-range cruise missiles. Early in a conflict, the Chinese would strike U.S. and partner-nation air bases, air and missile defenses, and command centers with large ballistic and cruise missile raids. The missiles would be armed with a mix of submunitions optimized for runway attack or destruction of aircraft on parking ramps. Numerous studies have demonstrated the vulnerability of airfields to such attacks.[187] As noted in the previous chapter, RAND played a large role in explaining Chinese strategy to U.S. audiences and in convincing the USAF and

---

[186] Jim Mattis, *Summary of the 2018 National Defense Strategy of The United States of America: Sharpening the American Military's Competitive Edge*, Washington, D.C.: U.S. Department of Defense, 2018, p. 4.

[187] See, for example, Stillion and Orletsky, 1999, and Heginbotham et al., 2015.

DoD to take this threat seriously.[188] RAND reports, wargames, and briefings provided strong evidence that the Chinese threat to USAF bases was significant and growing rapidly.

The Russian threat to USAF bases in Europe is growing as well. In any conflict (for example, in the Baltics), Russian doctrine would call for attacks on NATO air bases with manned aircraft, ballistic missiles, cruise missiles, and (in some circumstances) nuclear weapons. Although the Russian missile force is smaller and less advanced than the Chinese force, it nevertheless represents a significant threat to NATO air operations. NATO air bases do, however, have advantages that mitigate the threat somewhat. First, in contrast to the western Pacific, there are a large number of high-quality airfields available throughout western Europe, all of which are accessible by road and, in many cases, rail. Second, NATO fighter bases all have hardened shelters for fighters, and many have other hardened facilities. The primary problem for NATO is that larger aircraft, such as tankers, ISR platforms, and bombers, would need to operate from more remote bases to avoid the worst of these threats.

This historically anomalous era of sanctuary lasted almost 20 years before Chinese advances in long-range precision strike (primarily high-quality ballistic missiles) and battle networks (ISR and C2) advanced to the point that they threatened U.S. bases in East Asia. Although some authors, such as Stillion, Orletsky, and Stokes, had warned of the risk of missile attack as early as the 1990s, it wasn't until 2008 that, as noted in the previous chapter, the USAF began to take the threat seriously.[189]

Overview of RAND Research on Air Base Defense and Attack During This Period

The 2010–2020 period began with renewed DoD and USAF interest in air base defense. China's growing A2/AD capabilities were increasingly recognized as a threat to U.S. power projection capabilities and a harbinger of the end of unchallenged American military dominance. As noted in the previous chapter, the fortuitous appointment of Pacific Vision veterans Thomas Ehrhard and David Ochmanek to senior USAF and Office of the Secretary of Defense (OSD) positions meant that two highly respected defense analysts were well placed to educate DoD personnel and leaders on the growing missile threat to air bases. During this period, perhaps the most passionate advocate for improvements in air resiliency was Carl Rehberg. Rehberg was a retired USAF colonel who served in the USAF Quadrennial Defense Review office and, later, as

---

[188] Notably the 2007 RAND report *Entering the Dragon's Lair: Chinese Antiaccess Strategies and Their Implications for the United States* (Cliff et al., 2007).

[189] Andrew Hoehn and co-authors also highlighted the threat of ballistic missiles to forward bases more generally (not just in Asia) in a 2007 RAND report on force planning (Andrew R. Hoehn, Adam Grissom, David A. Ochmanek, David A. Shlapak, and Alan J. Vick, *A New Division of Labor: Meeting America's Security Challenges Beyond Iraq*, Santa Monica, Calif.: RAND Corporation, MG-499-AF, 2007).

the head of a USAF team focused on understanding and mitigating Chinese military advances.[190] Rehberg was the point of contact for multiple A2/AD-related PAF projects during these years, working closely with RAND researchers and making many substantive contributions to these studies.

The increased DoD and USAF interest in air base defense created a growing appetite for research and analysis, resulting in the commissioning of numerous RAND studies for DoD clients. Beginning in 2011, the USAF commissioned PAF to investigate "the combat support requirements and resource investments needed to support operations in denied environments, to include dispersed operations and other mitigation strategies."[191] Led by Robert Tripp, the Combat Operations in Denied Environments (CODE) studies represented RAND's largest analytical effort on air base defense since the Cold War and perhaps the most sustained body of work on the topic in RAND's history. Between 2011 and 2019, CODE project teams produced 20 reports on various aspects of air base defense. All but four of these reports were sponsored by the USAF.[192]

An earlier and ongoing body of work on Agile Combat Support (ACS) proved foundational to the work that Tripp and team conducted under the auspices of the CODE studies. This work, much of it led by Tripp, Don Snyder, or Patrick Mills, dates back to RAND research conducted for the USAF during the 1990s. As the USAF was increasingly asked to deploy relatively small rotational forces to the Middle East and Balkans for peace keeping and other duties, it found that a host of policies, practices, and capabilities designed for operations from MOBs were ill-suited for this new expeditionary environment. Responding to this demand, the USAF created the Expeditionary Aerospace Force and began to rethink its combat support concepts for this new environment and, in the process, commissioned a series of RAND studies assessing the relative attractiveness of various combat support options.[193]

---

[190] Now retired from civil service, Rehberg continues to make contributions as an adjunct staff member at the Center for Strategic and Budgetary Assessments. See, for example, Carl Rehberg and Mark Gunzinger, *Air and Missile Defense at a Crossroads: New Concepts and Technologies to Defend America's Overseas Bases*, Washington, D.C.: Center for Strategic and Budgetary Assessments, 2018.

[191] Brent Thomas, Mahyar A. Amouzegar, Rachel Costello, Robert A. Guffey, Andrew Karode, Christopher Lynch, Kristin F. Lynch, Ken Munson, Chad J. R. Ohlandt, Daniel M. Romano, Ricardo Sanchez, Robert S. Tripp, and Joseph V. Vesely, *Project AIR FORCE Modeling Capabilities for Support of Combat Operations in Denied Environments*, Santa Monica, Calif.: RAND Corporation, RR-427-AF, 2015, p. iii.

[192] As shown in Appendix A, Figure A.10, the most prolific authors on ABD between 2010 and 2020 were all associated with the CODE studies: Rachel Costello, Dahlia Goldfeld, Jacob Heim, Andrew Karode, Chris Lynch, Dan Romano, Brent Thomas, and Robert Tripp.

[193] Two important early contributions to this body of work were Robert S. Tripp, Lionel Galway, Paul S. Killingsworth, Eric Peltz, Timothy L. Ramey, John G. Drew, *Supporting Expeditionary Aerospace Forces: An Integrated Agile Combat Support Planning Framework*, Santa Monica, Calif.: RAND Corporation, MR-1056-AF, 1999; and Paul Killingsworth, Lionel A. Galway, Eiichi Kamiya, Brian Nichiporuk, Robert S. Tripp, and James C. Wendt, *Flexbasing: Achieving Global Presence for Expeditionary Aerospace Forces*, Santa Monica, Calif.: RAND Corporation, MR-1113-AF, 2000. For an overview of RAND's history of logistics research for the USAF, see Robert S. Tripp, *The Line Between Disorder and Order: Reflections on RAND's Role in The Evolution of Air Force*

This line of research and analysis ultimately led to the creation of the Lean-START (Lean Strategic Tool for the Analysis of Required Transportation) model, which underpins the analysis in Mills et al.'s *Balancing Agile Combat Support Manpower to Better Meet the Future Security Environment*.[194] Although Lean-START was not explicitly designed to address air base vulnerability, it was motivated by a desire to increase posture resilience and was developed at a time when the USAF was, once again, thinking seriously about the value of dispersing aircraft on and across bases as part of a portfolio of actions to counter growing standoff missile threats. One of the major impediments to a more distributed posture was that USAF force structure, personnel policies, and combat support concepts were still (despite the Expeditionary Aerospace Force) primarily designed to support a relatively small number of large bases. Lean-START was used by RAND analysts and the USAF to identify these constraints to (1) determine the limits of dispersing the force given current force structure and capabilities and (2) identify where ACS resources need to go to enable greater dispersal.[195] Mills and his research team had all either been involved in many years of RAND analysis on creating a more agile combat support system, worked on air base defense projects, or done both.

Four models provided the foundation for the CODE studies. "Together these models help illuminate combat support requirements, vulnerabilities, resiliency, and capability trade-offs and enable decisions concerning force posture, current and future investments, and theater-shaping strategies."[196] Two of the models (START and ROBOT [RAND Overseas Basing Optimization Tool]) were developed by prior PAF studies, while the other two (TAB-VAM [Theater Air Base Vulnerability Assessment Model] and TAB-ROM [Theater Air Base Resiliency Optimization Model]) were developed by the CODE team. START estimates manpower and equipment requirements for a variety of basing options, while ROBOT identifies least-cost allocations of war reserve materiel and transportation requirements.[197] TAB-VAM is a Monte Carlo simulation used "to analyze the complex trade-offs among basing strategies and threat mitigation options.

Logistics Thought, Santa Monica, Calif.: RAND Corporation, RR-3131-AF, 2020. For a history of the USAF EAF concept, see Richard G. Davis, *Anatomy of a Reform: The Expeditionary Aerospace Force*, Washington, D.C.: Air Force History and Museums Program, 2003.

[194] Patrick Mills, John G. Drew, John A. Ausink, Daniel M. Romano, and Rachel Costello, *Balancing Agile Combat Support Manpower to Better Meet the Future Security Environment*, Santa Monica, Calif.: RAND Corporation, RR-337-AF, 2014.

[195] Thanks to reviewer Mike Lostumbo for helping us understand the importance of RR-337 and the Lean-START model for air base resilience work generally and as part of the CODE effort specifically. Most of the observations in this paragraph are drawn from an email exchange between Lostumbo and co-author Vick, dated October 26 and 28, 2020.

[196] Thomas et al., 2015, p. iii.

[197] For more on the START model, see Don Snyder and Patrick Mills, *Supporting Air and Space Expeditionary Forces: A Methodology for Determining Air Force Deployment Requirements*, Santa Monica, Calif.: RAND Corporation, MG-176-AF, 2004. For more on ROBOT, see Mahyar A. Amouzegar, Ronald G. McGarvey, Robert S. Tripp, Louis Luangkesorn, Thomas Lang, and Charles Robert Roll, Jr., *Evaluation of Options for Overseas Combat Support Basing*, Santa Monica, Calif.: RAND Corporation, MG-421-AF, 2006.

The model allows the user to assess and compare a wide range of scenarios, aircraft beddowns, base recovery capabilities, infrastructure investments, passive and active missile defense options, and concepts of operation."[198] The final model, TAB-ROM, "searches the entire space of user-defined enemy attack strategies for a given scenario and finds the most cost-effective way to improve Blue sortie generation through investments in active missile defense, hardened aircraft shelters, fuel storage, and/or ADR [airfield damage repair]."[199]

Although CODE reports represented roughly half of the ABD/A reports published between 2010 and 2020, there were another 20 or so smaller studies that made important contributions as well.

In 2015, a team led by Michael Lostumbo produced an influential report (sponsored by OSD) that quantified the benefits of dispersing on and across bases, as well as the survivability/sortie generation trades associated with conducting air operations at great distances.[200] The team also developed the concept of "cluster basing," in which a group of airfields would be located in close proximity so that they could be mutually supporting and sit underneath a single air/missile defense umbrella. Lostumbo and project member Jacob Heim would go on to lead several other major air base defense studies during this time period and participate in many OSD or USAF sponsored conferences, wargames, and other planning activities related to air base defense.

In 2016, a team led by Forrest Morgan produced a report describing an alternative framework for air base resiliency analysis. The Operational Resilience Analysis Model (ORAM), based on sequential game theory, sought to better represent the actions of adaptive enemies to blue strategies. The "operational resilience analysis was a multistage process in which we ran multiple cases through ORAM and adjusted each side's strategies based on what the outputs of those runs revealed."[201] After multiple iterations of the model (in which each side is adapting to actions' by the other), a set of robust Blue resilience improvements could be identified. A high-level cost analysis of these improvements would then be made to estimate the cost of that particular portfolio. With its game theoretic foundation, the ORAM study offered a complementary approach to the CODE projects. As the CODE projects evolved, the analytical approach was refined to better capture choices by adaptive adversaries, and the ORAM effort was discontinued.

Table 7.1 displays the most common analytical and policy objectives of RAND projects during this time period. Almost 70 percent of the reports were focused on one of two objectives:

---

[198] Thomas et al., 2015, p. xiv.

[199] Thomas et al., 2015, p. xv.

[200] Report authors were Michael Lostumbo, David Frelinger, Jacob Heim, Brian Jackson, Amelia Becker, Stephen Worman, and Paul Dreyer.

[201] Jeff Hagen, Forrest E. Morgan, Jacob L. Heim, and Matthew Carroll, *The Foundations of Operational Resilience—Assessing the Ability to Operate in an Anti-Access/Area Denial (A2/AD) Environment: The Analytical Framework, Lexicon, and Characteristics of the Operational Resilience Analysis Model (ORAM)*, Santa Monica, Calif.: RAND Corporation, RR-1265-AF, 2016, p. xiii.

conduct air operations during and after conventional attack or identifying the most effective means to generate sorties under heavy attack. Seven of the reports were focused on the best means to defend airfields. Another four sought to quantify the effects of attacks. Of these, two were focused on better understanding weapons effects, specifically small submunition damage mechanisms against aircraft parked in the open. The other two quantified the broader effects of missile attacks on PACAF bases.

Table 7.1. Top Policy/Analytical Objectives of RAND Reports on Air Base Defense and Attack: 2010–2020

Policy/Analytical Objective	Number of RAND Reports
Conduct air operations during and after conventional attack	19
Identify most effective means to generate sorties under heavy attack	11
Identify most effective means to defend airfields	7
Identify most effective means to attack and defend airfields	3
Quantify effects of missile attacks on PACAF bases	2
Quantify effect of tactical ballistic missile/submunition attacks on parked aircraft	2

NOTE: The table only includes reports in the top six categories, not all reports in the decade.

A Sampling of Reports

The 46 RAND reports published between 2010 and November 2020 were overwhelmingly focused on technical and operational problems. The studies documented in these reports primarily used operations research techniques, typically including sophisticated computer models and simulations. Unfortunately, most of these 48 reports are not available to the public. Two of the five reports discussed below are, however, representative of the larger body of RAND work: *The U.S.-China Military Scorecard* and *Building Agile Combat Support Competencies to Enable Evolving Basing Concepts* use analytical techniques that are similar to the other air base defense studies.

- *The U.S.-China Military Scorecard: Force, Geography, and the Evolving Balance of Power, 1996–2017* (2015)
- *Air Base Attacks and Defensive Counters: Historical Lessons and Future Challenges* (2015)
- *Distributed Operations in a Contested Environment: Implications for USAF Force Presentation* (2019)
- *Air Base Defense: Rethinking Army and Air Force Roles and Functions* (2020)
- *Building Agile Combat Support Competencies to Enable Evolving Adaptive Basing Concepts* (2020).

The first report, *The U.S.-China Military Scorecard: Force, Geography, and the Evolving Balance of Power, 1996–2017*, was published in 2015. The study sought to understand how the U.S.-China military balance was changing. The research team examined each country's military capabilities in ten operational areas and produced a scorecard for each area for four years: 1996, 2003, 2010, and 2017 (projected). The scorecards included four metrics for air and missile capabilities, two for maritime forces, and four for space, cyber, and nuclear capabilities.[202] "By employing a consistent methodology, the scorecards provide a portrait of trends over time. To provide insight into the impact of geography and distance, each of the scorecards evaluates capabilities in the context of two scenarios: a Taiwan invasion and a Spratly Islands campaign."[203] The scorecards of greatest interest for our purposes are "Scorecard 1: Chinese Capability to Attack Air Bases" and "Scorecard 4: U.S. Capability to Attack Chinese Air Bases."

Regarding China's ability to attack U.S. air bases, the authors, based on their modeling of missile attacks on Kadena Air Base in Japan, conclude "that even a relatively small number of accurate missiles could shut the base to flight operations for critical days at the outset of hostilities, and focused, committed attacks might close a single base for weeks." The report also considers the implications of a future Chinese IRBM, concluding that "with an inventory of just 50 IRBMs, China could keep Andersen AFB closed to large aircraft for more than eight days."[204] After noting the benefits of improved active defenses, hardened aircraft shelters, improved runway repair and aircraft dispersal, the authors, nonetheless, concluded that "the growing number and variety of Chinese missiles will almost certainly challenge the U.S. ability to operate from forward bases" and that "basing issues will greatly complicate U.S. efforts to gain air superiority over the battlefield."[205]

The authors see the U.S. ability to attack Chinese air bases as dependent on the survivability of manned aircraft and availability of all-weather precision weapons. They "modeled attacks on the 40 Chinese air bases within unrefueled fighter range of Taiwan, and, separately, on the smaller number from which Chinese aircraft could range the Spratly Islands."[206] For the Taiwan scenario, the model results saw an improving picture for the United States, with runway attacks closing Chinese runways for an average of eight hours in 1996 but between two and three days for 2010 through 2017. The Spratly Islands scenario was even more favorable for the United States; "In all four snapshot years, U.S. air forces could effectively close all of China's air bases opposite the Spratly Islands for the first week of operations."[207]

The *U.S.-China Military Scorecards* report is unique in its scope and depth. There is no other analysis in the public domain remotely as detailed or as comprehensive. The closest analogue would be the Shlapak et al. reports on a Taiwan-China conflict published in 2000 and 2009, but both of those efforts were relatively small, foundation-funded studies. *Scorecards* was USAF-

---

[202] Heginbotham et al., 2015, pp. xxi–xxii.

[203] Heginbotham et al., 2015, p. xix.

[204] Heginbotham et al., 2015, p. 64.

[205] Heginbotham et al., 2015, p. xxiii.

[206] Heginbotham et al., 2015, p. xxv.

[207] Heginbotham et al., 2015, p. xxv.

sponsored and better-resourced, benefiting from the efforts of 14 researchers with a wide range of subject-matter expertise and diverse modeling skills. It remains the most definitive treatment of the topic available in the public literature.

Air Base Attacks and Defensive Counters: Historical Lessons and Future Challenges, also published in 2015, sought to place ABD/A in historic context. Its key findings were that (1) air base attacks have been a common feature in both minor and major conflicts occurring between 1914 and 2014; (2) the primary (and enduring) components of air base defense were first identified during World War I and include active defense, CC&D, hardening, dispersal on and off base, and postattack recovery; (3) after the Cold War ended, the United States developed a new way of war that took for granted rear area sanctuary; (4) emerging long-range strike capabilities are bringing this era of sanctuary to an end; and (5) a combination of measures will be needed to counter emerging threats—the specific mix will vary depending on regional geography, access, adversary capabilities, and U.S. objectives.

The report recommended that USAF and DoD planners (1) consider the air base, the airspace above and near it, and the surrounding land as a battlespace, a place where defenders cannot expect sanctuary; (2) develop new concepts for deployment to and operation of air bases under attack; and (3) explore organizational options to better support distributed and dispersed operations, with a focus on whether the typical USAF fighter wing with three squadrons operating at no more than two locations is appropriate for this new environment.[208]

Distributed Operations in a Contested Environment: Implications for USAF Force Presentation, published in 2019, squarely tackled the last recommendation of the 2015 report, namely that alternative organizational options should be considered to enhance air base resiliency. Specifically, the report "identifies capabilities the Air Force needs to carry out distributed operations in a contested environment. It then assesses whether the current force presentation model can provide these capabilities and how it compares with alternative models."[209]

The report's key findings were (1) the USAF force presentation model and operating concepts are based on assumptions that are incompatible with a contested environment; (2) the contested environment will force the USAF to trade efficiency for survivability; (3) developing concepts for distributed operations will require close collaboration between operations and combat support communities, and (4) more analysis is needed for command and control, support, and other implications of distributed operations for nonfighter forces.

The report recommends that the USAF (1) determine resource and access requirements for distributed operations; (2) simulate heavy air, missile, and ground attacks in home station

---

[208] Vick, 2015, pp. xii–xv.

[209] Miranda Priebe, Alan J. Vick, Jacob L. Heim, Meagan L. Smith, *Distributed Operations in a Contested Environment: Implications for USAF Force Presentation*, Santa Monica, Calif.: RAND Corporation, RR-2959-AF, 2019, p. iii.

training and exercises; (3) consider creating integrated base defense units; (4) hold regular exercises that include communication disruptions; (5) cross-train airmen to reduce the personnel demands of distributed operations; (6) consider the possible role of the group in distributed operations before eliminating the peacetime Group echelon; and (7) use exercises and additional analysis to explore force presentation implications of distributed operations.[210]

Air Base Defense: Rethinking Army and Air Force Roles and Functions, published in 2020 and commissioned by Headquarters U.S. Air Force Europe (USAFE), addressed the gap between the growing threat to USAFE air bases and shortfalls in joint ground-based air defense systems. The threat of greatest concern to USAFE leaders was cruise missiles, but the study considered the full range of threats, from swarming drones to hypersonic missiles. The study broadly assessed threats, defense options, and roles and functions constraints in order to identify seven alternative courses of action for consideration by USAF leaders. The strengths and weaknesses of these prospective courses of action were then assessed to determine whether they were likely to address the fundamental roles and functions issues.

The report key findings were (1) air base defense, against both ground and air threats, has been an enduring area of disagreement and frustration for the Army and USAF; (2) although many factors are at play, the misalignment of service responsibilities and priorities for air base defense is hindering the correction of enduring shortfalls; (3) limitations of joint force development processes, Army resource constraints, and USAF ambivalence have also contributed to an air base defense roles and functions roadblock; (4) the USAF may be able to bypass the roles and functions roadblock through innovation and use of advanced technologies such as directed energy; and (5) the most robust strategy to improve air base defenses would pursue parallel lines of effort.

The report makes three recommendations: (1) demonstrate institutional commitment to air base defense by funding and advocating for substantial enhancements in capability areas already assigned to the USAF such as security forces and passive defense programs; (2) use the USAF culture of innovation to break down roles and functions barriers; and (3) propose a new memorandum of understanding with the Army to establish ground-based air defense of air bases as a USAF responsibility.[211]

Building Agile Combat Support Competencies to Enable Evolving Adaptive Basing Concepts was also published in 2020. The authors define adaptive basing (AB) as "the U.S. Air Force's effort to extend the survivability of combat forces and the operational resilience of those forces in CDO [contested and degraded, and operationally limited] environments through combinations of traditional and adaptive strategies that could vary from site to site and campaign to

---

[210] Priebe et al., 2019, pp. x–xiii.

[211] Vick et al., 2020, pp. xiii–xiv.

campaign."[212] The report identifies three ACS competencies that would be foundational if the USAF were to adopt adaptive basing: integrated basing (networks of bases), flexible operations, and rapid scalability. Among the many contributions of this research was the creation of four "archetypes" of bases for analysis: Dispersal bases, Stay and Fight bases, Temporary bases, and Traditional bases. This framework has proven highly useful both in this analysis and other RAND studies.[213] This report findings included several implications for the ACS community and USAF, including (1) "The design of current force packages is ill-suited for executing AB concepts," (2) "Implementing AB concepts would require the ACS community to develop new competencies for employing ACS capabilities," and (3) "AB represents a fundamental pivot in how the Air Force presents forces to warfighting commands."[214] The study recommended that (1) "The ACS community should consider overhauling the force packages used for deploying and presenting forces to combatant commanders," (2) "The ACS community should consider personnel skill design and personnel development activities that could help fulfill ACS requirements for AB," and (3) "The Air Force should consider an experimentation campaign to test various aspects of implementing AB concepts."[215]

RAND Contributions

RAND ABD/A research was clearly ahead of the curve in the 1990s. Stillion and Orletsky illustrated how GPS and other guidance technologies would make cruise and ballistic missiles a vastly more lethal threat to air bases, and Shlapak and Vick showed the strategic significance of ground threats to air bases. RAND teams in the 2000s continued to be thought leaders, especially the two Shlapak studies that integrated air base attack into campaign analysis and in the 13th Air Force study that supported the Pacific Vision game, as well as the efforts by RAND staff in educating USAF and other DoD personnel and leaders regarding the emerging missile threat to bases.

In the 2010–2020 period, RAND built on this recent foundation of research and analysis, especially in the development of increasingly sophisticated analytical techniques. Table 7.2 highlights some of the contributions made by RAND during this period.

---

[212] Patrick Mills, James A. Leftwich, John G. Drew, Daniel P. Felten, Josh Girardini, John P. Godges, Michael J. Lostumbo, Anu Narayanan, Kristin Van Abel, Jonathan William Welburn, and Anna Jean Wirth, *Building Agile Combat Support Competencies to Enable Evolving Adaptive Basing Concepts*, Santa Monica, Calif.: RAND Corporation, RR-4200-AF, 2020, p. x.

[213] For example, Priebe et al. (2019) used the (then in development) Mills basing framework in their analysis and report.

[214] Mills et al., 2019, p. xv.

[215] Mills et al., 2019, pp. xv–xvii.

Table 7.2. RAND Contributions to Research on Air Base Defense and Attack, 2010–2020

RAND Contribution	Documented in
Major advances in ABD/A analytical methods (TAB-VAM, TAB-ROM, Lean-START, ORAM [Operational Resilience Analysis Model], CATAPULTA [Covert and Aerial Threat Analysis Program to Understand the Lethality of Targeting of Airbases])	CODE team reports (2011–2019); Mills et al. (2014); Hagen et al. (2016); Savitz et al. (2015)
Implications of standoff threats for USAF overseas posture	Lostumbo et al. (2013); Vick and Heim (2013)
Comprehensive open-source assessment of U.S.-China military balance, including relative ABD/A capabilities	Heginbotham et al. (2015)
Lessons learned from air base attacks during 26 conflicts	Vick (2015)
Implications of distributed air operations for force presentation concepts	Priebe et al. (2019)
Implications of adaptive basing concepts for combat support	Mills et al. (2014); Mills et al. (2020); Wirth et al. (2020)

The CODE studies, likely the largest and most sustained ABD/A analysis in RAND's history, represent RAND's single greatest ABD/A accomplishment during this period, and they have made significant contributions to analytical methods as well as policy development. This research has

> directly informed, and in some cases significantly reshaped, the decisions
> through which DoD seeks to enhance its capabilities, including investment
> requests for base operating support resources, requirements for expeditionary
> medical capabilities and infrastructure, requirements for active and passive
> defense, and new concepts for adaptive basing. In January 2015, this work was
> highlighted by the Air Force's Advanced Capability and Deterrence Panel
> (ACDP) as a hallmark of innovation in resiliency for the Pacific.[216]

Heginbotham et al.'s *China Military Scorecards* study was ambitious in scope and depth, modeling force interactions across air, sea, and space. As of late 2020, it is one of a kind, the only publicly available quantitative analysis of trends in the U.S.-China military balance. The 2015 Vick report, although quite small compared with the other efforts, offered a comprehensive historical treatment of air base attacks, including estimates for aircraft lost on the ground across 26 conflicts occurring between 1914 and 2014. The report offered both historical lessons and a look at future challenges, including a chapter arguing that the American way of war developed after the Cold War was dependent on rear area sanctuary that could no longer be counted on. The report also included a chapter arguing that Chinese missile developments were intended to be a disruptive innovation that would overturn U.S. conventional power projection dominance. Finally, the 2019 Priebe et al. report delved into the organizational requirements of distributed

---

[216] RAND Corporation, *Fostering Innovation in the Defense Department: Examples from RAND's Federally Funded Research and Development Centers*, Santa Monica, Calif.: RAND Corporation, CP-852, June 2016, p. 3.

operations, arguing that the group echelon was uniquely valuable for command and control of distributed operations during periods of communication disruption.

The next and final chapter offers a brief summary of the 70-year arc of RAND research and analysis on ABD/A and highlights a few key findings.

8. Conclusions

This report sought to capture some of the breadth and depth of RAND research on air base defense and attack over the past 70 years and place it within the context of an ever-evolving geopolitical, military, and technological landscape. RAND responded to these changes in strategic demand and policy priorities with alacrity and urgency but, given its charter, was not entirely bound by them. At critical junctures, RAND led its DoD and USAF sponsors, identifying emerging threats to air bases and potential solutions well before the broader community acknowledged them.

RAND did not discover the fundamental concepts of air base attack or defense. Airmen from multiple nations were primarily responsible for developing air base attack concepts, although missile force planners, special operators, and insurgents have all made their unique contributions. Regarding defenses, both active and passive approaches were first used by air base defenders over 100 years ago during World War I, and the state of the defender's art advanced greatly during World War II. Concepts such as sheltering aircraft in hardened structures; dispersing aircraft on and across bases; runway repair; camouflage, concealment, and deception; and the protection and dispersal of fuels and munitions all came out of the crucible of war—not postwar analysis.

Within these broad categories, however, RAND did design a variety of solutions to the problems of ABD/A, including unconventional ideas, such as using elevators to lower bombers into hardened shelters (1957) and using loitering RPVs to enable air base attacks by manned aircraft (1971). RAND engineers also helped the USAF design and test an advanced runway attack munition (1966). But RAND's greatest contributions were in its disciplined and creative application of more formal analytical tools to the problem of air base defense and attack. RAND invented and applied these tools so that the relative utility of various offensive and defensive concepts could be measured in a systematic way.

Most significant of these analytical techniques was RAND's practice of combining cost and effectiveness metrics into an integrated analysis. RAND studies helped identify which of a given set of options were most effective and affordable, although computational and resource limits typically constrained how many options might be considered in a single analysis. For example, until relatively recently, RAND studies did not assess trade-offs between active and passive defenses at the base level.[217]

---

[217] In addition to computational limitations, there are constraints on the number of variables that even contemporary analyses can assess. RAND studies were also influenced by service roles and missions. The Army was assigned the responsibility of providing point air defenses for fixed facilities (including air bases), so USAF-sponsored studies typically did not consider these programs that were outside of the USAF functional assignments. Rather, RAND studies on air defenses were overwhelmingly focused on early warning, C2, and defensive counter-air operations—

RAND analysis increasingly showed that rarely was the choice simply between two distinct options. Rather, the question was what mix of capabilities were most cost-effective to defeat a given threat or to put the adversary's air bases out of action. Most recently, advanced simulation, modeling, and statistical techniques have made it possible for RAND researchers to ask what mix of capabilities is most cost-effective and robust to defend against a wider range of adversary "attack vectors" (i.e., their strategies and operational concepts).

RAND's early reports on ABD/A were written at the beginning of the nuclear age, when little was known about nuclear effects and policymakers lacked an agreed-upon way of thinking or talking about the role of nuclear weapons in national strategy. In the 1950s, RAND made seminal contributions to the nation's understanding of both the tactical/technical issues and the geopolitical/strategic problems facing the nation. Several RAND analysts who contributed to this work became quite famous after leaving RAND (e.g., Herman Kahn, Albert Wohlstetter, Bernard Brodie, Andrew Marshall), and one (Thomas Schelling) received a Nobel Prize for his work on game theory.

Building on the foundational 1950s analyses of nuclear threats to air bases, RAND researchers then expanded this body of work over the following decades to include defense against conventional attacks, the most effective means to attack runways and hardened structures on enemy air bases, assessing ground force threats to air bases, and, finally, the challenges posed by adversary acquisition of highly accurate long-range ballistic and cruise missiles.

Most of the RAND reports produced on this topic remain unavailable to the public. Consequently, many of the most interesting research findings and recommendations cannot be discussed in this document. We can, however, identify some of the broader insights stemming from this large body of work. We summarize these below.

Key Findings

Air Bases Have Always Been, and Are Likely to Remain, Priority Targets in Wars

This is due to two essential facts. First, modern air power has proven to be a versatile and essential element of military power, one that, at minimum, must be countered to prevail in conflict. Second, unlike navies and armies, which generate combat power from the fleet at sea and maneuver forces in the field, air forces generate combat power from fixed airfields. As long as air power retains these characteristics—that it is both vital and tethered to vulnerable bases—adversaries will have great incentives to attack airfields. RAND research on the history of attacks on air bases demonstrates the historic truth of these observations; in 26 conflicts spanning over 100 years, combatants have attacked airfields on thousands of occasions.

---

all functions assigned to the USAF. RAND conducted many studies on area air defense but usually as separate efforts. Thus, active defenses are somewhat underrepresented in the RAND ABD/A report collection.

Air Base Attackers Will Rarely Limit Themselves to a Single Attack Mode

The early 1950s were an exception in the way that strategic nuclear competition was constrained by delivery platforms; only long-range bombers could carry the large nuclear (gravity) bombs of the period. This lasted until the early 1960s, when ICBMs and SLBMs broadened the attack options available.

In conventional wars, combatants have attacked airfields with aircraft, cruise missiles, naval gunfire, artillery, mortars, rockets, commando raids, armored forces, and drones. Most combatants—even insurgents—have multiple options for attacking airfields and will use them as conditions dictate. It is often preferable to combine attack modes and vectors to create synergies among offensive weapon systems (e.g., precursor cruise missile attacks on BMD radars followed by ballistic missile attacks) and to distract and confuse defenders.

As of 2020, hypersonic missiles and swarming small drones offer additional attack vectors, further complicating the defender's problem.

There Are No Simple or Cheap Means to Defend Air Bases

A review of RAND research findings and the longer history of ABD/A offers no panacea solution to the problem of airfield vulnerability. No broad category (passive or active defense) offers perfect protection, nor is either category consistently the most cost-effective option. Similarly, no single type of active defense (e.g., fighter interceptors versus ground-based SAMs) or passive defense (e.g., hardened shelters versus dispersal) offers either complete protection or is reliably the most cost-effective.[218] There also are trade-offs among competing goals. For example, if the USAF wanted to protect fighter aircraft at forward bases from rockets, mortars, sniper rifles, anti-tank guided weapons, small drones, submunitions, and near misses from larger air-delivered munitions, it might choose to build concrete shelters similar to those built at USAF bases during the Cold War. But such shelters would offer no protection from an advanced cruise missiles, are expensive, would take years to build, and would be limited to those bases where the USAF had an enduring presence. In contrast, simple defensive structures designed to defeat small submunitions could be produced cheaply, deployed rapidly, and erected on relatively short notice anywhere in the world. Yet, light, deployable shelters (often only offering overhead

---

[218] Regarding RAND research findings, we are aware of no single analysis that included a detailed analysis of all the major weapon options for attack and all the major defensive options. The analytical tools available in the 1950s and 1960s were sufficiently limited that the best a study could do was to consider a range of defensive options against a given threat or a few offensive options against a few targets at an airfield. In following decades, advances in modeling and simulation greatly expanded the breadth and depth of options considered but practical constraints (time, money, manpower) always forced some choices between breadth and depth of analysis. Even with today's analytical tools this is the case. The CODE studies have come closest to realizing this goal but even they face similar constraints in available programmer time and, most problematic, the risk of "blowing up the model" as the number of variables and interactions among variables grows. For a thoughtful treatment of these analytical challenges, see Paul K. Davis, *Analysis to Inform Defense Planning Despite Austerity*, Santa Monica, Calif.: RAND Corporation, RR-482-OSD, 2014.

protection) would offer limited or no protection from direct hits from unitary weapons, drone attack, direct fire weapons (e.g., sniper fire), or near misses from larger munitions.

Neither active nor passive defenses are inherently cheaper or preferable when employed at scale. For these reasons, past combatants and most current militaries combine approaches whenever feasible, seeking out the most cost-effective combination of tools to solve the specific problem they face. Past RAND analyses have helped sponsors identify portfolios of capabilities that are cost-effective for a bounded problem (specified threat or range of threats, specified set of options, estimated costs, and specified performance metric), but they cannot definitely say that one option is always superior.

Aircraft Dispersal on and Across Bases Has Renewed Salience for Air Base Defense

As noted in multiple places in the report, dispersing aircraft on and across bases has been a key component of air base defense portfolios since World War I. Aircraft dispersal arguably achieved maximum salience in the 1950s, when the vulnerability of the U.S. bomber force to nuclear attack became one of the nation's most urgent defense problems. Although not cost-free, dispersal was an option that could be implemented relatively quickly, certainly compared with building nuclear-hardened shelters or deploying active defenses at every base. It is no coincidence that the second RAND report on the topic of ABD/A was on the costs associated with dispersing B-36 wings.[219] As the standoff missile threat increased in the 2010–2020 period, various concepts for distributed operations have again regained prominence as among the most versatile and executable. As discussed in Chapter 7, distributed operations present a host of challenges for the USAF but, on the whole, are often easier to implement than other passive defense options and do not require massive investments in infrastructure at bases that may not be needed in the next war.

Air Base Defense and Attack Are Best Understood from a Systems Perspective

As suggested above, under conditions of uncertainty, planners must assess the performance of a range of defensive options against an even wider range of enemy offensive options. Systems analysis can help in multiple ways. First, it helps the planner understand how key air base processes work. Similar to an industrial facility, an air base takes inputs (e.g., aircraft, personnel, fuel, munitions), then follows formal procedures and protocols (e.g., mission planning, aircraft maintenance, fueling, arming) to create usable products (aircraft manned and ready for missions) that can be measured using output metrics (e.g., sorties generated, enemy aircraft shot down per mission, targets struck).

The great diversity of possible enemy attack vectors, their effects on the air base system, and the many defensive options available greatly complicate defensive planning. Making matters

---

[219] RAND Corporation, Cost Analysis Section, 1952.

worse, the relative cost and effectiveness of attacking and defending options may vary greatly. Thus, defensive options must be understood in terms of their cost-effectiveness as well as their robustness across enemy attack vectors. From its very beginning, RAND has applied the tools of systems analysis to the ABD/A problem. Advances in modeling and simulation are allowing RAND's analysts to look at many more attack and defense combinations than was previously possible and with much greater detail, providing insights into how defensive portfolios perform on various metrics against a wide range of adversary attack strategies. These methodological advances also make possible consideration of complex, hybrid attacks, where the attacker uses missiles, aircraft, ground forces, drones, and other weapons/forces in combination.

Looking to the Future

The vulnerability of forward air bases to attacks from long range precision weapons is leading airmen to rethink the relationship between air power and the air base. Alternatives under consideration include distributed operations (with no more than 12–18 aircraft per location); moving in and out of airfields to confound adversary targeting; use of cheap, easily deployable shelters; abandoning heavily damaged airfields; and using longer-range aircraft and standoff weapons to operate outside the worst threat rings. Additionally, RAND analysts are working with USAF sponsors on concepts to reduce dependence on forward air bases through the use of large numbers of low-cost unmanned aircraft that can be launched and recovered without runways.[220] Another possibility is a greater dependence on short takeoff and vertical landing (STOVL) aircraft, such as the Marine Corps' F-35B. These reduce the dependence on major air bases but at great cost in range and payload. Finally, with the abrogation of the Intermediate Nuclear Force (INF) Treaty, the United States now has the option of deploying conventionally armed land-based cruise or ballistic missiles with ranges up to 5,500 km. Although the USAF is not currently exploring these latter options, mobile land-based cruise missiles offer yet another way to reduce air power's dependence on fixed air bases.

Whatever the technologies and concepts eventually embraced, the USAF is likely to field some combination of the above options to ensure that air power remains viable in contested environments. At the same time, we can be sure that adversaries will develop new weapons and concepts to find and attack these air power assets.

"The Battle of the Airfields" will likely look quite different in the coming decades but, if history is any guide, RAND will continue to be actively involved in supporting USAF and DoD efforts to ensure the resiliency of American air power—whether that air power comprises unmanned aircraft launched from trucks, mobile missiles, or manned aircraft flying from more traditional air bases.

---

[220] See Thomas Hamilton and David Ochmanek, *Operating Low-Cost, Reusable, Unmanned Aerial Vehicles in Contested Environments: Preliminary Evaluation of Operational Concepts*, Santa Monica, Calif.: RAND Corporation, RR-4407-AF, 2020.

Appendix A. Statistics on Report Authors and Co-Author Networks

In this appendix, we identify the authors who were instrumental in producing the body of RAND reports on ABD/A. Whereas most of our analysis up to this point has been dedicated to identifying key themes and reports, here we look at the most prolific contributors and how they are connected to one another as co-authors of reports. (Appendix B looks separately at first authorship.)

As noted above, first authorship is not the sole indicator of meaningful contributions. We can also look at total appearances as an author, whether as first author or elsewhere on the byline. Figure A.1 lists the top 30 authors from 1951 to 2020, ranked by total contributions as an author in descending, clockwise order. The circles include the number of publications per author, and the lines show co-author connections among individuals in the group. Thicker lines represent more frequent collaborations. For example, there is a strong connecting line between Robert Guffey and Rachel Costello, who have been on eight reports together. Also note that some of the most prolific authors (Tripp, Karode, Lynch, Heim, Costello, Thomas, Romano, Guffey, and Goldfeld) were core members of the Combat Operations in Denied Environments (CODE) research team. Tripp oversaw this body of work with Heim, Costello, Goldfeld, and Thomas leading several efforts under the overall CODE umbrella. The vast majority of RAND air base defense reports completed between 2010 and 2020 came from this stream of analysis. Unfortunately, only one of these reports is publicly available (RR-427 on CODE models), and, therefore, the CODE analyses are not discussed in this report. Michael Lostumbo (nine reports co-authored) was another research leader in this field, conducting major studies for various DoD clients between 2010 and 2020. As is the case with the CODE work, we are not able to discuss the important contributions of Lostumbo and his co-authors. Finally, Mills (seven reports co-authored) has played an important and growing role in CODE and related basing studies.

Figure A.1. Top Authors and Their Co-Authorship Connections, 1951–2020

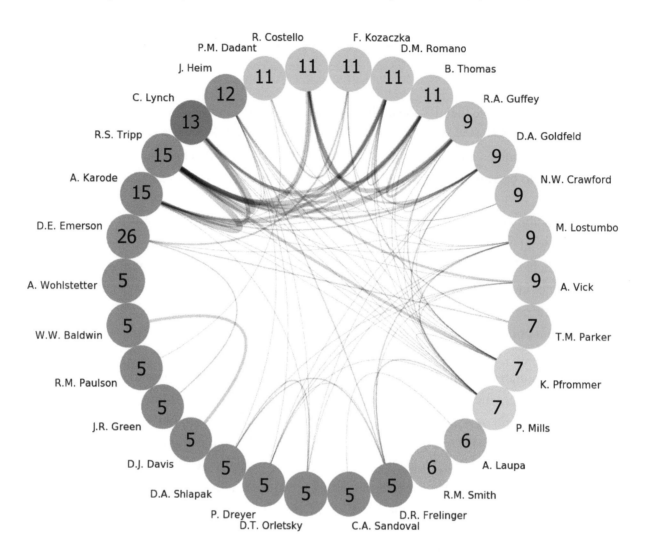

Similar plots are included for the top authors in each major era (Figures A.4, A.6, A.8, A.10). In addition to these plots, we include network graphs showing all co-authorship connections (Figures A.5, A.7, A.9, A.11). The network graphs make it easier to see the clusters of authors who worked more closely together. In these graphs, we also identify the authors who have the highest *degree centrality* and *betweenness centrality*. In graph theory, centrality is a way of measuring the influence of individual nodes. Nodes (authors) with more edges (co-authorships) have higher degree centrality. Nodes that connect other nodes by the shortest path have higher betweenness centrality. For our purposes, authors with high betweenness can be thought of as those who act as the bridge between large clusters of other authors.

Figure A.2. Network Graph of All Authors, 1951–2020

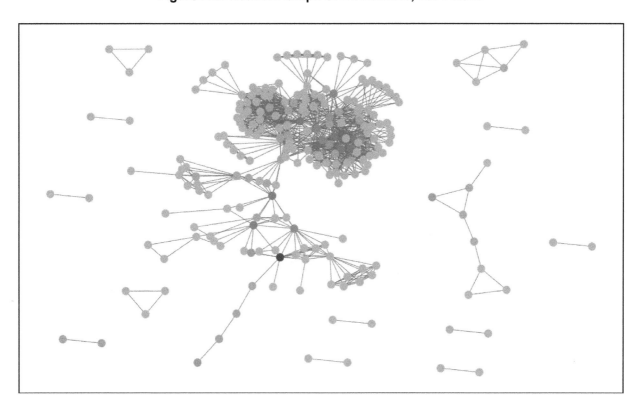

NOTES: Darker blue circles are authors with more publications. Authors without co-author connections (sole authors) are not included.

When viewing Figures A.4–A.11, note that graphs are much sparser in earlier years. This is primarily because it was the convention at the time for RAND reports to have fewer co-authors. Figure A.3 shows the average number of authors per report by time period. It went from an average of approximately one author per report in the 1950s and 1960s to six in the 2010s.

Figure A.3. Average Number of Authors per Report (with standard deviation bars)

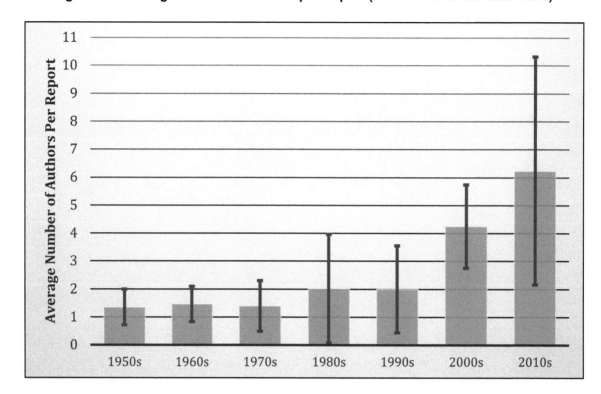

Figure A.4. Top Authors, 1951–1969

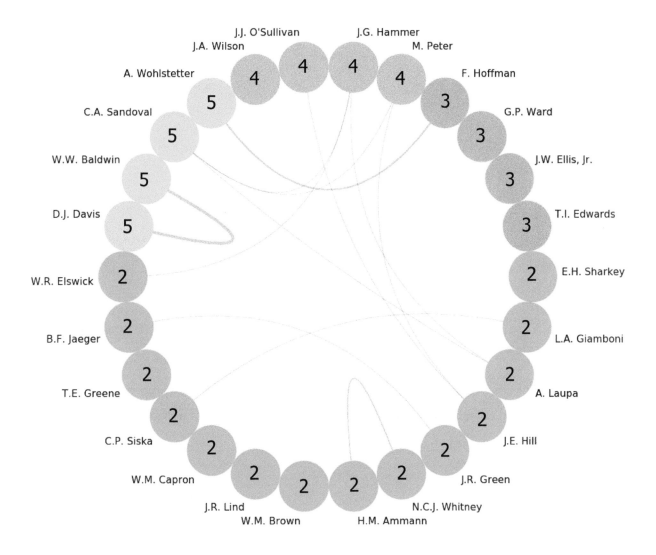

Figure A.5. Network Graph of 1951–1969, Including 42 Authors and 35 Co-Author Connections

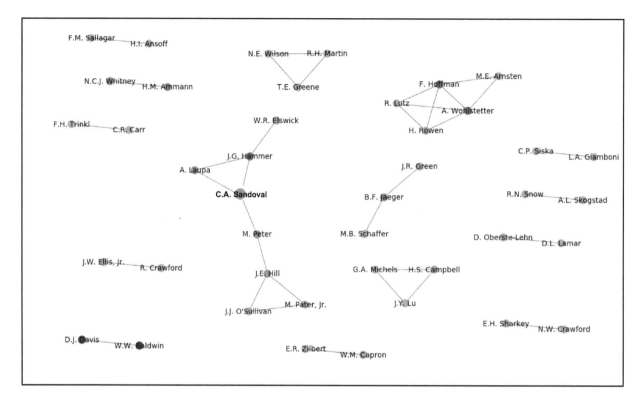

NOTES: Darker blue = more reports. Red = highest degree centrality. Green outline = highest betweenness centrality.

Figure A.6. Top Authors, 1970–1989

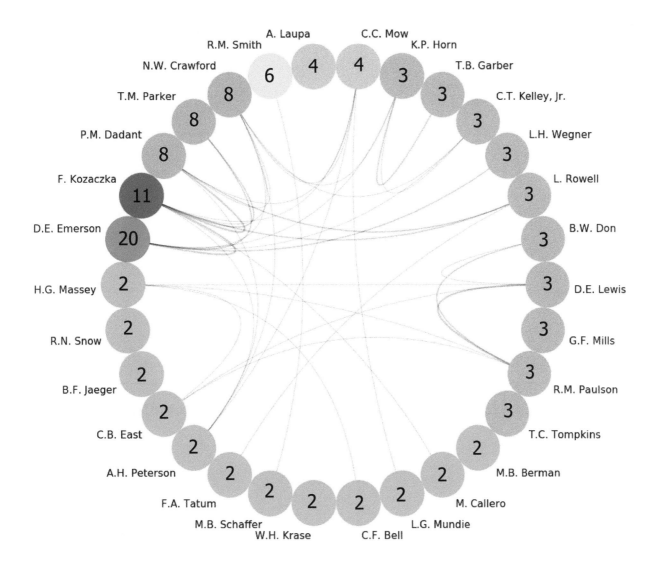

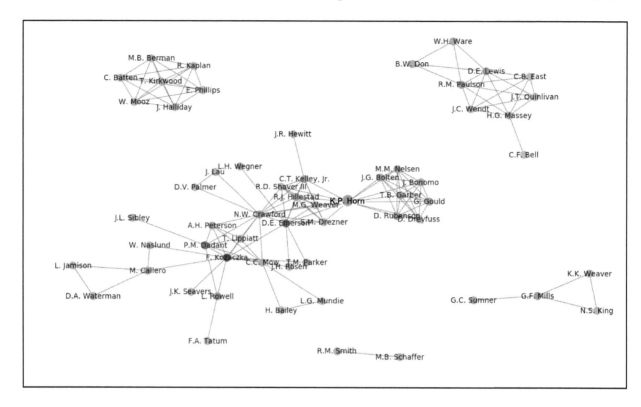

NOTES: Darker blue = more reports. Red = highest degree centrality. Green outline = highest betweenness centrality.

Figure A.8. Top Authors, 1990–2009

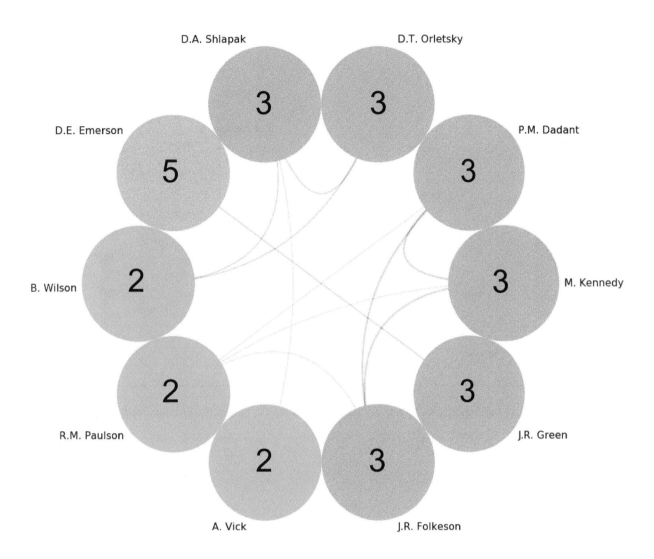

Figure A.9. Network Graph of 1990–2009, Including 26 Authors and 53 Co-Author Connections

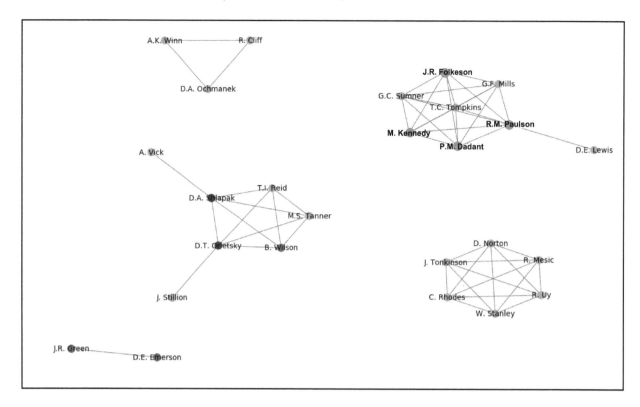

NOTES: Darker blue = more reports. Red = highest degree centrality. No betweenness centrality due to low number of connections.

Figure A.10. Top Authors, 2010–2020

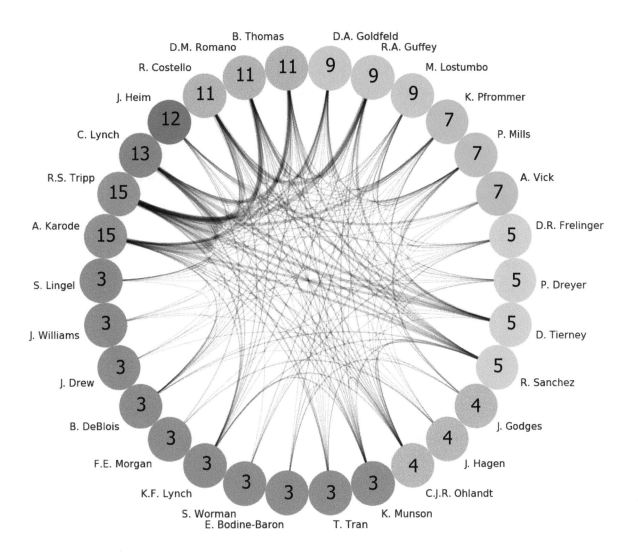

Figure A.11. Network Graph of 2010–2020, Including 121 Authors and 877 Co-Author Connections

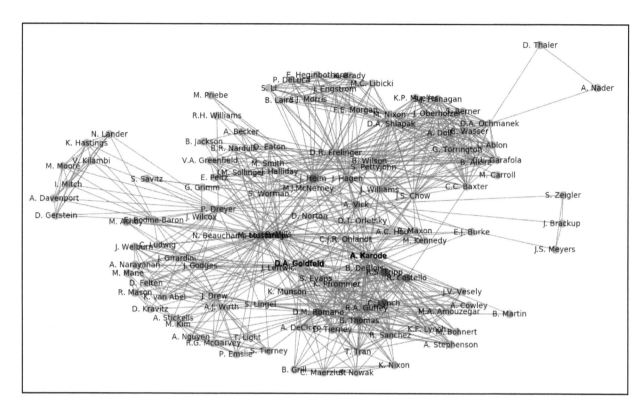

NOTES: Darker blue = more reports. Red = highest degree centrality. Green outline = highest betweenness centrality.

Appendix B. First Authors by Decade

This appendix provides a series of figures showing the number of reports produced by the top first authors in each decade. This is in contrast to Appendix A, which shows the authors with the greatest number of reports, regardless of whether they were first author or not. Because the 1990s and 2000s had so few reports, the figures show the number of publications produced by all first authors. For all other decades, the figures only show first authors with two or more publications.

Figure B.1 shows a count of the top ten first-author contributors from our dataset between 1951 and 2020. Donald E. Emerson, creator of the TSAR and TSARINA models that were discussed in Chapter 4, had 25 publications as first author. Phillip Dadant, Natalie Crawford, and Ted Parker were also quite active with nine, seven, and six first author credits, respectively. The number of first authored publications is, of course, an imperfect metric. Albert Wohlstetter, who was hugely influential and the lead author of the single most famous RAND report on basing, was one of six RAND analysts with five first author credits on basing reports.[221]

---

[221] See Chapters 3, 4, and 5, respectively, for more information about each author's contributions.

Figure B.1. Researchers with the Most Publications as First Author, 1951–2020

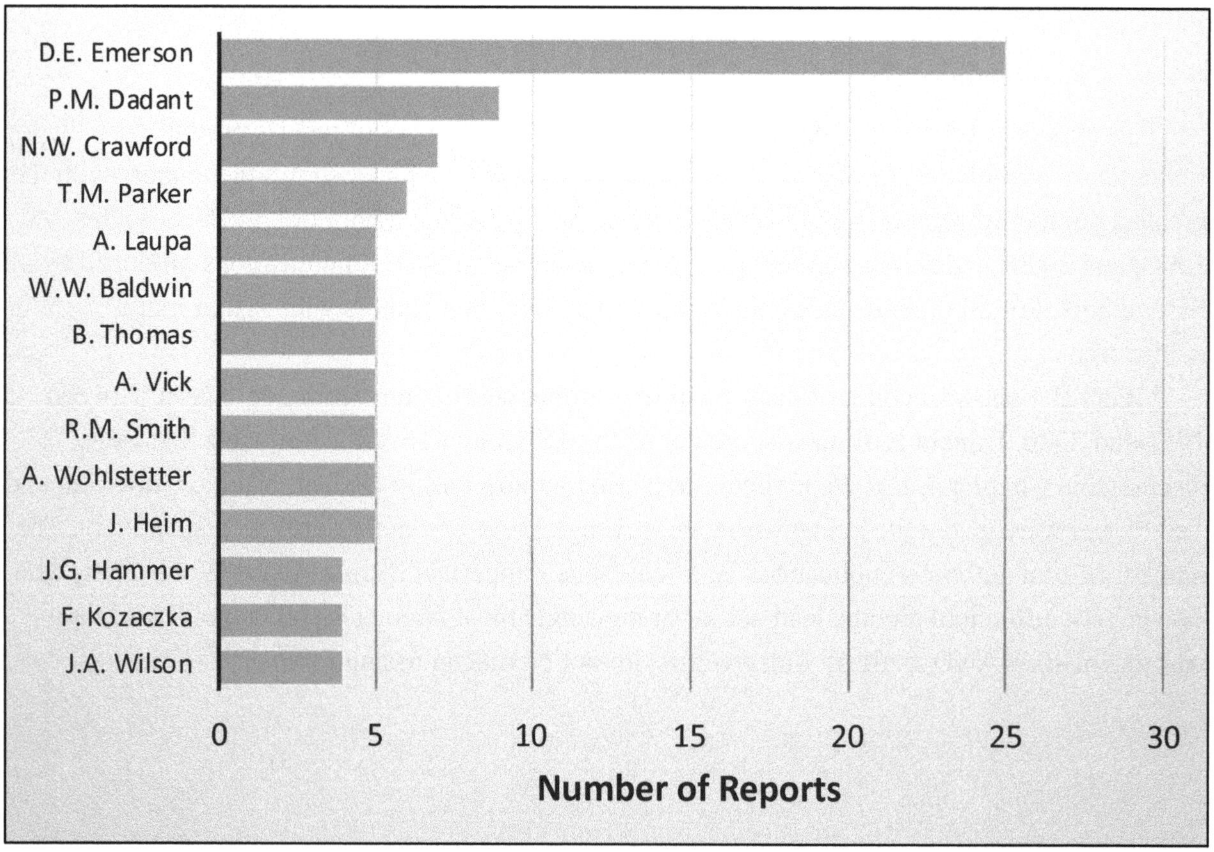

NOTE: Count limited to reports on air base defense or attack.

Figure B.2. Top First Authors in the 1950s

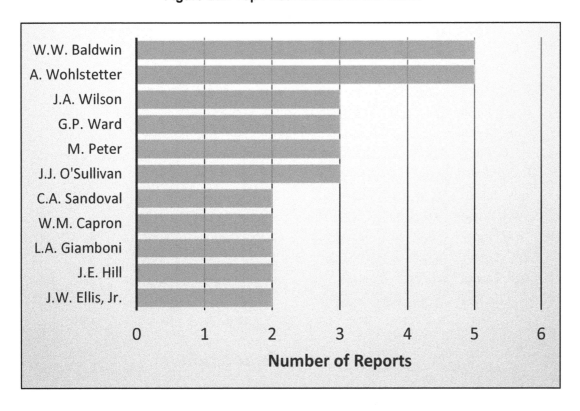

Figure B.3. Top First Authors in the 1960s

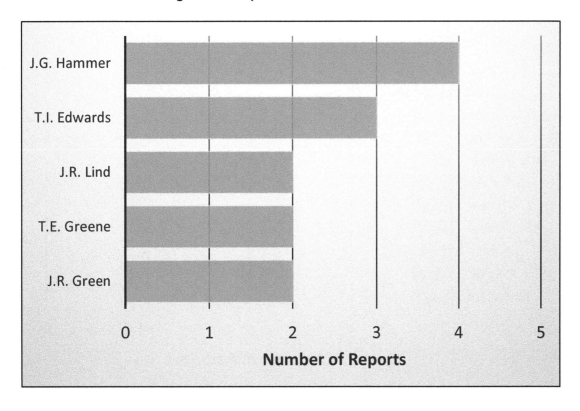

Figure B.4. Top First Authors in the 1970s

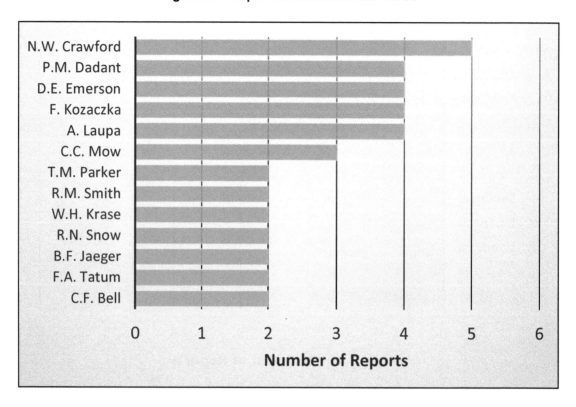

Figure B.5. Top First Authors in the 1980s

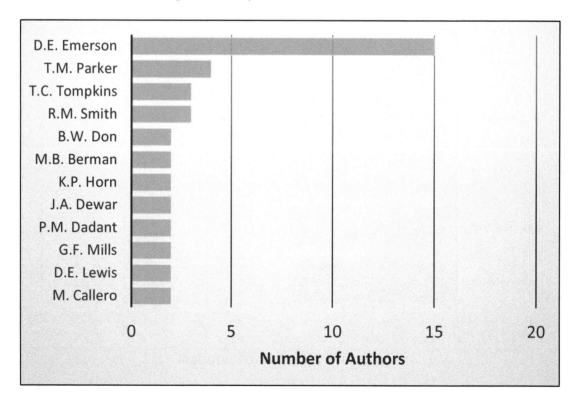

Figure B.6. Top First Authors in the 1990s

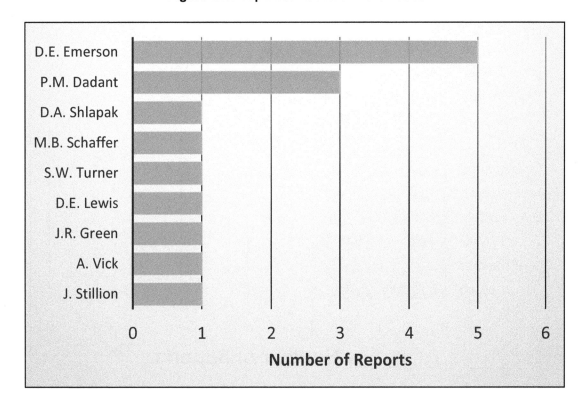

Figure B.7. Top First Authors in the 2000s

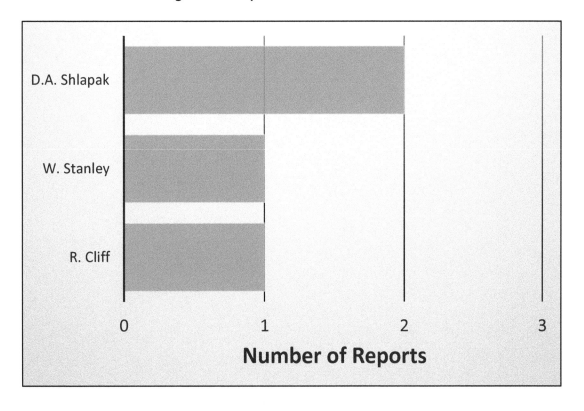

Figure B.8. Top First Authors in the 2010s

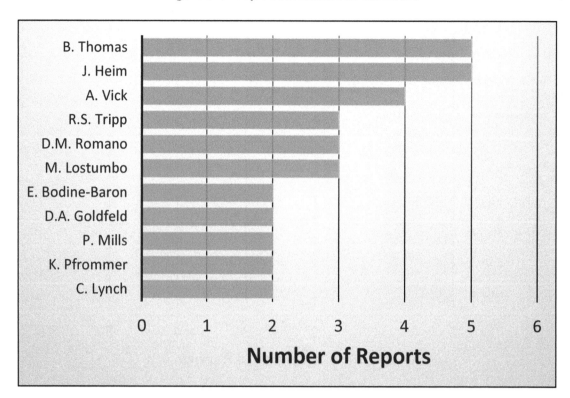

Appendix C. Chronological Listing of Publicly Available RAND Reports on Air Base Attack or Defense

This appendix lists RAND publications that are discussed in this report.[222] Many of these reports were not available to the public prior to this project. We have provided a URL for each report where available; some reports were still in the process of being posted online when this report was published, but all should be available at https://www.rand.org/pubs.html in early 2021.

Reports beginning with "D," "IN," "WN," or "WD" were previously restricted to internal use only. For previously classified reports, declassification dates are provided where available. Note that the number of publications varies greatly by year; in some years, there were no reports published on air base defense or attack, so the listing skips some years (e.g., 1962 and 1964). Also note that first names were not available in RAND records for all authors. Where available, first names are shown in brackets in the full bibliography (which follows this appendix).

1951

Wohlstetter, Albert, *Economic and Strategic Considerations in Air Base Location: A Preliminary Review*, D-1114, https://www.rand.org/pubs/documents/D1114.html.

1952

RAND Corporation, Cost Analysis Section, *The Cost of Decreasing Vulnerability of Air Bases by Dispersal: Dispersing a B-36 Wing*, R-235, https://www.rand.org/pubs/reports/R235.html. (Declassified by Air Force Declassification Office on April 18, 2016.)

1953

RAND staff, *Special Staff Report: The Selection of Strategic Air Bases*, R-244-S. (Classification canceled by USAF on July 26, 1963. Air Force Declassification Office reviewed and confirmed declassification on August 10, 2015.)

Jackson, Victor G., *Use of Alternative Recovery Bases by Air Defense Interceptor Aircraft*, RM-1152. (Declassified by USAF on February 1, 1961.)

---

[222] For a more complete listing of RAND and external references used in this report, see the bibliography following this appendix.

1954

Baldwin, W. W., and D. J. Davis, *Preservation of Tactical Air Combat Potential in Western Europe: Wing-Level Defense Against A-Bombing*, RM-1462. (Declassified by Air Force Declassification Office on October 24, 2017.)

Skogstad, A. L., R. N. Snow, *Preservation of Tactical Air Combat Potential in Western Europe: Allied Force, Base, and Radar Structure*, RM-1310. (Declassified by Air Force Declassification Office on August 13, 1965.)

Stockton, A. C., *Preservation of Tactical Air Combat Potential in Western Europe: Underground Hangars for Fighter-Bomber Operations*, RM-1230.

Tuck, R. E., Preservation *of Tactical Air Combat Potential in Western Europe: Guided Missile Defense Potential*, RM-1312. (Declassified by Air Force Declassification Office on April 18, 2016.)

Wohlstetter, Albert, and Fred Hoffman, *Defending a Strategic Force After 1960*, D-2270.

Wohlstetter, Albert, Fred Hoffman, and Michael E. Arnsten, *Measures to Protect Airbase Bulk Fuel Stocks*, RM-1398.

Wohlstetter, A. J., F. S. Hoffman, R. J. Lutz, and H. S. Rowen, *Selection and Use of Strategic Air Bases*, R-266.

1955

Peter, Marc, Jr., *Estimates of Aircraft Shelter Costs*, D-3347-PR.

Ward, G. P., *B-52 Main and Satellite Base Installations Cost: Warm and Cold Climate*, D-2908-PR.

Ward, G. P., *SAC Dispersal: Estimated Installations Costs of Wing, Main and Satellite Bomber Bases, Hard and Soft*, D-3005-PR.

Ward, G. P., *SAC Dispersal: Estimated Costs of Recovery Bases and Home Shelters*, D-2966-1-PR.

1956

Hill, J. E., J. J. O'Sullivan, and Marc Peter, Jr., *Restoration of Runways Following Attack*, D-3573-PR.

Peter, Marc, Jr., and Charles Sandoval, *Variations in Aircraft Shelter Costs*, D-3509-PR.

RAND Staff Report, *Protecting U.S. Power to Strike Back in the 1950s and 1960s*, R-290. (Declassified by Air Force Declassification Office on April 18, 2016.)

1957

McGlothlin, W. H., *A Method for Studying the SAC Base Dispersal Problem*, D(L)-4740.

Sandoval, Charles A., *Size and Cost of Shelter Doors Subjected to Long Duration Nuclear Blast Loading*, D-4630.

1958

Attaway, L. D., *Programmed Warning of U.S.-Based SAC Against Air-Breathing Threats: Problems and Suggested Solutions*, RM-2236. (Declassified by Air Force Declassification Office on October 23, 2018.)

O'Sullivan, J. J., *Time to Clear Building Debris from Air Base Runways*, D(L)-5032.

Sandoval, Charles A., *A Handbook for Estimating Material Requirements and Costs of Shelter Doors Subjected to Long-Duration Blast Loading*, RM-2277.

Wakeley, Jay T., *Dispersal: An Expedient for Protecting SAC*, (annotated briefing), S-113.

1959

Amman and Whitney, Consulting Engineers, *Hardened Alert Hangars*, D(L)-6570.

Brown, William M., *Vulnerability of Quick-Reacting Sheltered Missiles and Aircraft During Launch*, D-6625.

Capron, William M., *Let's Build a B-52 Shelter Now*, D-6159-PR.

Levine, Robert A., *Hard Homes for Heavy-Bomber Crews*, D-6744-PR.

Sandoval, Charles A., *Report of a Meeting Concerning B-52 Hardening*, D-6894-PR.

Weidlinger, Paul, *An Alert Shelter for the B-58*, D-6698-PR.

1960

Lamar, D. L., and D. Oberste-Lehn, *Operation Cliff Dweller: Determination of Site Location for Hardened Aircraft Bases*, D-8078.

Schelling, Thomas C., *Sitting Ducks or Decoys: The High Cost of SAC Dispersal to Large-City Airfields*, D-7329.

Tamplin, A. R., *Operation Cliff Dweller: Hardening Bases, Atmospheric Control and Disease*, D-8020.

1961

DeWeerd, Harvey A., *Bases for V-Bombers in the USA: A Cold War Proposal*, D-8472.

Hammer, J. G., *Representative Costs for B-52 Protective Alert Shelters*, D-8974.

Hammer, J. G. and Armas Laupa, *On Hardened Basing of B-52 Aircraft*, D-9513.

Hammer, J. G., and C. A. Sandoval, *Comparison and Evaluation of Protective Alert Shelter Concepts for SAC Aircraft*, D-8740, https://www.rand.org/pubs/documents/D8740.html.

1963

Green, J. R., *The Performance of G.P. Bombs in Penetrating and Cratering Concrete Runways*, D-11110-PR.

Jaeger, B. F., M. B. Schaffer, *Tentative Thoughts on Non-Nuclear IRBM's for Attacking Parked Aircraft*, D(L)-11285-PR, https://www.rand.org/pubs/documents/D11285.html.

1965

Edwards, T. I., *Demonstration of Emergency Repair of Bomb-Cratered Runway*, D-13633-PR.

1966

Crawford, Robert, and J. W. Ellis Jr., *Airbase Defense and Security with Application to Thailand*, D-15350-ARPA/AGILE.

Edwards, T. I., *Successful Tests of a RAND-Recommended Runway Cratering Device*, D-15075.

Hammer, J. G., and W. R. Elswick, *Conventional Missile Attacks Against Aircraft on Airfields and Aircraft Carriers*, RM-4718-PR. (Declassified by Air Force Declassification Office on October 23, 2018.) https://www.rand.org/pubs/research_memoranda/RM4718.html.

Wilson, N. E., and B. F. Jaeger, *Runbst-Runway Cutting III: Runway Cutting Programs*, D-15197-PR.

1969

Sharkey, E. H., *Some Hand-Done Calculations on Attacks Against Runways*, D-18821-PR.

1970

Burkholz, Gail M., *Aircraft Payload Limits for the Airbase Attack Study*, D-20322-PR. (Declassified on January 20, 1977.)

Dadant, Phillip M., *Report of Trips to Brief Airbase Attack Study: Europe—7 to 11 September 1970; Langley, Pentagon and Andrews—19 to 22 October 1970; West Coast Study Facility—6 November 1970*, IN-21293-PR.

Laupa, Armas, *Time Estimates to Repair Cratered Runways*, D-19937-PR. (Declassified on January 20, 1977.)

Sharkey, E. H., *Estimated CEP of Weapon Delivery for Attacks Against Aircraft Shelters*, D-20250-PR. (Declassified on January 20, 1977.)

1971

Kozaczka, Felix, and J. K. Seavers, *Examination of Warsaw Pact Airbase Attack Capability Against Unsheltered Aircraft in the Central Region of NATO*, IN-21822-PR. (Declassified on December 31, 1979.)

Krase, W. H., *Target-Marking Systems for RPVs Used as Designator Vehicles for Airbase Attack*, IN-21646-PR. (Declassified on April 21, 1978.)

Krase, W. H., *Vehicle Estimates for Target Designation RPVs*, IN-21682-PR. (Declassified on April 21, 1978.)

1973

Dadant, Phillip M., *Measures of Effectiveness and the TALLY/TOTEM Methodology*, P-5062.

Kozaczka, Felix, *Modeling Nuclear Vulnerability of NATO Tactical Airfields*, IN-22741-DDPAE.

1974

Farquhar, Peter H., *Recommendations for Improving AHAB: An Interactive Decision Aid for Tactical Commanders*, IN-23132-PR.

Kozaczka, Felix, *Nuclear Vulnerability of NATO Tactical Airfields: Some Calculations of Nuclear Radiation Exposure*, IN-22933-DDPAE. (Declassified on January 3, 1983.)

Neu, Carl Richard, *Attacking Hardened Air Bases (AHAB): A Decision Analysis Aid for the Tactical Commander*, R-1422-PR, https://www.rand.org/pubs/reports/R1422.html.

Tatum, F. A. and L. N. Rowell, *PROBE I: A Differential Equation Model for Comparing Fighter Escort and Airbase Attack Systems in a Counter-Air Operation*, R-1413-PR, https://www.rand.org/pubs/reports/R1413.html.

1975

Snow, R. N., *An Effectiveness Model for Multiple Attacks Against an Airbase Area Complex*, R-1639-PR, https://www.rand.org/pubs/reports/R1639.html.

1976

Emerson, Donald E., *AIDA: An Airbase Damage Assessment Model*, R-1872-PR, https://www.rand.org/pubs/reports/R1872.html.

1982

Callero, Monti, Lewis Jamison, and D. A. Waterman, *TATR: An Expert Aid for Tactical Air Targeting*, N-1796-ARPA, https://www.rand.org/pubs/notes/N1796.html.

Emerson, Donald E., *USAFE Airbase Operations in a Wartime Environment*, P-6810, https://www.rand.org/pubs/papers/P6810.html.

Emerson, Donald E., *An Introduction to the TSAR Simulation Program: Model Features and Logic*, R-2584-AF, https://www.rand.org/pubs/reports/R2584.html.

Emerson, Donald E., *TSAR User's Manual: Volume I—Program Features, Logic, and Interactions*, N-1820-AF, https://www.rand.org/pubs/notes/N1820.html.

Emerson, Donald E., *TSAR User's Manual: Volume II—Data Input, Program Operation and Redimensioning, and Sample Problem*, N-1821-AF, https://www.rand.org/pubs/notes/N1821.html.

Emerson, Donald E., *TSAR User's Manual: Volume III—Variable and Array Definitions, and Other Program Aids for the User*, N-1822-AF, https://www.rand.org/pubs/notes/N1822.html.

Wegner, Louis H., *The Taxiway Repair Schedule Problem: A Heuristic Rule and a Branch-and-Bound Solution*, N-1883-AF, https://www.rand.org/pubs/notes/N1883.html.

1983

Don, Bruce W., *A Cost Effectiveness Analysis of First Wave OCA Attacks Using Ground Launched Missiles*, IN-24937-AF.

1986

Lewis, Donald E., Bruce W. Don, Robert M. Paulson, and Willis H. Ware, *A Perspective on the USAFE Collocated Operating Base System*, N-2366-AF.

1987

Halliday, John M., *Tactical Dispersal of Fighter Aircraft: Risk, Uncertainty, and Policy Recommendations*, N-2443-AF.

Lansdowne, Zachary F., *Computing the Closure Probability for an Airfield Defended with Anti-Tactical Ballistic Missiles*, WD-3651-A/ACQ.

Tompkins, Thomas C., *An Airbase Ground Attack Scenario in Central Europe*, WD-3549-AF.

1988

Don, Bruce W., Donald E. Lewis, Robert M. Paulson, and Willis H. Ware, *Survivability Issues and USAFE Policy*, N-2579-AF.

Mills, Gary F., and Gerald C. Sumner, *Costs of Airbase Survivability Options*, IN-25488-AF.

1995

Shlapak, David A., and Alan Vick, *Check Six Begins on the Ground: Responding to the Evolving Ground Threat to U.S. Air Force Bases*, MR-606-AF, www.rand.org/t/MR606.

Vick, Alan J., *Snakes in the Eagle's Nest: A History of Ground Attacks on Air Bases*, MR-553-AF, www.rand.org/t/MR553.

1999

Stillion, John, and David T. Orletsky, *Airbase Vulnerability to Conventional Cruise-Missile and Ballistic-Missile Attacks: Technology, Scenarios, and U.S. Air Force Responses*, MR-1028-AF, www.rand.org/t/MR1028.

2000

Shlapak, David A., David T. Orletsky, and Barry Wilson, *Dire Strait? Military Aspects of the China-Taiwan Confrontation and Options for U.S. Policy*, MR-1217-SRF, www.rand.org/t/MR1217.

2009

Shlapak, David A., David T. Orletsky, Toy I. Reid, Murray Scot Tanner, and Barry Wilson, *A Question of Balance: Political Context and Military Aspects of the China-Taiwan Dispute*, MG-888-SRF, www.rand.org/t/MG888.

2013

Lostumbo, Michael J., Michael J. McNerney, Eric Peltz, Derek Eaton, David R. Frelinger, Victoria A. Greenfield, John Halliday, Patrick Mills, Bruce R. Nardulli, Stacie L. Pettyjohn, Jerry M. Sollinger, and Stephen M. Worman, *Overseas Basing of U.S. Military Forces: An Assessment of Relative Costs and Strategic Benefits*, RR-201-OSD, www.rand.org/t/RR201.

2015

Thomas, Brent, Mahyar A. Amouzegar, Rachel Costello, Robert A. Guffey, Andrew Karode, Christopher Lynch, Kristin F. Lynch, Ken Munson, Chad J.R. Ohlandt, Daniel M. Romano, Ricardo Sanchez, Robert S. Tripp, and Joseph V. Vesely, *Project AIR FORCE Modeling Capabilities for Support of Combat Operations in Denied Environments*, RR-427-AF, www.rand.org/t/RR427.

Heginbotham, Eric, Michael Nixon, Forrest E. Morgan, Jacob L. Heim, Jeff Hagen, Sheng Li, Jeffrey Engstrom, Martin C. Libicki, Paul DeLuca, David A. Shlapak, David R. Frelinger, Burgess Laird, Kyle Brady, and Lyle J. Morris, *The U.S.-China Military Scorecard Force, Geography, and the Evolving Balance of Power, 1996–2017*, RR-392-AF, www.rand.org/t/RR392.

Vick, Alan J., *Air Base Attacks and Defensive Counters: Historical Lessons and Future Challenges*, RR-968-AF, www.rand.org/t/RR968.

2016

Hagen, Jeff, Forrest E. Morgan, Jacob L. Heim, and Matthew Carroll, *The Foundations of Operational Resilience—Assessing the Ability to Operate in an Anti-Access/Area Denial (A2/AD) Environment: The Analytical Framework, Lexicon, and Characteristics of the Operational Resilience Analysis Model (ORAM)*, RR-1265-AF, www.rand.org/t/RR1265.

2019

Priebe, Miranda, Alan Vick, Jacob Heim, and Meagan Smith, *Distributed Operations in a Contested Environment: Implications for USAF Force Presentation*, RR-2959-AF, www.rand.org/t/RR2959.

2020

Mills, Patrick, James A. Leftwich, John G. Drew, Daniel P. Felten, Josh Girardini, John P. Godges, Michael J. Lostumbo, Anu Narayanan, Kristin Van Abel, Jonathan William

Welburn, and Anna Jean Wirth, *Building Agile Combat Support Competencies to Enable Evolving Adaptive Basing Concepts*, RR-4200-AF, www.rand.org/t/RR4200.

Vick, Alan, Sean Zeigler, Julia Brackup, and John Speed Meyers, *Air Base Defense: Rethinking Army and Air Force Roles and Functions*, RR-4368-AF, www.rand.org/t/RR4368.

Wirth, Anna Jean, Thomas Light, Daniel M. Romano, Shane Tierney, Ronald G. McGarvey, Moon Kim, Michael J. Lostumbo, Amanda Nguyen, Paul Emslie, and John G. Drew, *Evaluating Alternative Maintenance Manpower Force Structure Concepts for the F-35A*, RR-4433-AF, www.rand.org/t/RR4433.

Bibliography

RAND Reports on Air Base Attack or Defense[223]

Amman and Whitney, Consulting Engineers, *Hardened Alert Hangars*, Santa Monica, Calif.: RAND Corporation, D(L)-6570, 1959.

Attaway, L[eland] D., *Programmed Warning of U.S.-Based SAC Against Air-Breathing Threats: Problems and Suggested Solutions*, Santa Monica, Calif.: RAND Corporation, RM-2236, 1958. (Declassified by Air Force Declassification Office on October 23, 2018.)

Baldwin, W[oodson] W., and D[avis] J. Davis, *Preservation of Tactical Air Combat Potential in Western Europe: Wing-Level Defense Against A-Bombing*, Santa Monica, Calif.: RAND Corporation, RM-1462, 1954. (Declassified by Air Force Declassification Office on October 24, 2017.)

Barlow, E. J., *RAND Air Defense Analysis*, Santa Monica, Calif.: RAND Corporation, RM-562, 1951. (Declassified by the Air Force Declassification Office on August 14, 2015.)

Brown, William M., *Vulnerability of Quick-Reacting Sheltered Missiles and Aircraft During Launch*, Santa Monica, Calif.: RAND Corporation, D-6625, 1959.

Burkholz, Gail M., *Aircraft Payload Limits for the Airbase Attack Study*, Santa Monica, Calif.: RAND Corporation, D-20322-PR, 1970. (Declassified on January 20, 1977.)

Callero, Monti, Lewis Jamison, and D[onald] A. Waterman, *TATR: An Expert Aid for Tactical Air Targeting*, Santa Monica, Calif.: RAND Corporation, N-1796-ARPA, 1982. As of December 9, 2020:
https://www.rand.org/pubs/notes/N1796.html

Capron, William M., *Let's Build a B-52 Shelter Now*, Santa Monica, Calif.: RAND Corporation, D-6159-PR, 1959.

Crawford, Robert, and J[ack] W. Ellis Jr., *Airbase Defense and Security with Application to Thailand*, Santa Monica, Calif.: RAND Corporation, D-15350-ARPA/AGILE, 1966.

---

[223] During RAND's first two decades, authors were usually listed with just first and middle initials and last name. First names are not available in RAND records for all authors. Where available, we show them in brackets. Also, at least one author (Thomas I. Edwards) used both conventions. Listings for Edwards's reports therefore show his name with full first name or with first initial, reflecting how it appeared on a given report.

Dadant, Phillip M., *Report of Trips to Brief Airbase Attack Study: Europe—7 to 11 September 1970; Langley, Pentagon and Andrews—19 to 22 October 1970; West Coast Study Facility—6 November 1970*, Santa Monica, Calif.: RAND Corporation, IN-21293-PR, 1970.

Dadant, Phillip M., *Measures of Effectiveness and the TALLY/TOTEM Methodology*, Santa Monica, Calif.: RAND Corporation, P-5062, 1973. As of December 9, 2020:
https://www.rand.org/pubs/papers/P5062.html

DeWeerd, Harvey A., *Bases for V-Bombers in the USA: A Cold War Proposal*, Santa Monica, Calif.: RAND Corporation, D-8472, 1961.

Don, Bruce W., *A Cost Effectiveness Analysis of First Wave OCA Attacks Using Ground Launched Missiles*, Santa Monica, Calif.: RAND Corporation, IN-24937-AF, 1983.

Don, Bruce W., Donald E. Lewis, Robert M. Paulson, and Willis W. Ware, *Survivability Issues and USAFE Policy*, Santa Monica, Calif.: RAND Corporation, N-2579-AF, 1988. As of December 9, 2020:
https://www.rand.org/pubs/notes/N2579.html

Edwards, Thomas I., *A Runway Interdiction Weapon and a Modular Fire Bomb: Development and Tests in Progress at Eglin AFB*, Santa Monica, Calif.: RAND Corporation, 1965, not available to the general public.

Edwards, T[homas] I., *Demonstration of Emergency Repair of Bomb-Cratered Runway*, Santa Monica, Calif.: RAND Corporation, D-13633-PR, 1965.

Edwards, Thomas I., *Successful Tests of a RAND-Recommended Runway Cratering Device*, Santa Monica, Calif.: RAND Corporation, D-15075, 1966.

Emerson, Donald E., *AIDA: An Airbase Damage Assessment Model*, Santa Monica, Calif.: RAND Corporation, R-1872-PR, 1976. As of December 9, 2020:
https://www.rand.org/pubs/reports/R1872.html

Emerson, Donald E., *USAFE Airbase Operations in a Wartime Environment*, Santa Monica, Calif.: RAND Corporation, P-6810, 1982. As of December 9, 2020:
https://www.rand.org/pubs/papers/P6810.html

Emerson, Donald E., *An Introduction to the TSAR Simulation Program: Model Features and Logic*, Santa Monica, Calif.: RAND Corporation, R-2584-AF, 1982. As of December 9, 2020:
https://www.rand.org/pubs/reports/R2584.html

Emerson, Donald E., *TSAR User's Manual: Volume I—Program Features, Logic, and Interactions*, Santa Monica, Calif.: RAND Corporation, N-1820-AF, 1982. As of December 9, 2020:
https://www.rand.org/pubs/notes/N1820.html

Emerson, Donald E., *TSAR User's Manual: Volume II—Data Input, Program Operation and Redimensioning, and Sample Problem*, Santa Monica, Calif.: RAND Corporation, N-1821-AF, 1982. As of December 9, 2020:
https://www.rand.org/pubs/notes/N1821.html

Emerson, Donald E., *TSAR User's Manual: Volume III—Variable and Array Definitions, and Other Program Aids for the User*, N-1822-AF, 1982. As of December 9, 2020:
https://www.rand.org/pubs/notes/N1822.html

Farquhar, Peter H., *Recommendations for Improving AHAB: An Interactive Decision Aid for Tactical Commanders*, Santa Monica, Calif.: RAND Corporation, IN-23132-PR, 1974.

Green, J[ohn] R., *The Performance of G.P. Bombs in Penetrating and Cratering Concrete Runways*, Santa Monica, Calif.: RAND Corporation, D-11110-PR, 1963.

Hagen, Jeff, Forrest E. Morgan, Jacob L. Heim, and Matthew Carroll, *The Foundations of Operational Resilience—Assessing the Ability to Operate in an Anti-Access/Area Denial (A2/AD) Environment: The Analytical Framework, Lexicon, and Characteristics of the Operational Resilience Analysis Model (ORAM)*, Santa Monica, Calif.: RAND Corporation, RR-1265-AF, 2016. As of December 9, 2020:
https://www.rand.org/pubs/research_reports/RR1265.html

Halliday, John M., *Tactical Dispersal of Fighter Aircraft: Risk, Uncertainty, and Policy Recommendations*, Santa Monica, Calif.: RAND Corporation, N-2443-AF, 1987. As of December 9, 2020:
https://www.rand.org/pubs/notes/N2443.html

Hammer, J[ohn] G., *Representative Costs for B-52 Protective Alert Shelters*, Santa Monica, Calif.: RAND Corporation, D-8974, 1961.

Hammer, J. G., and W[illiam] R. Elswick, *Conventional Missile Attacks Against Aircraft on Airfields and Aircraft Carriers*, Santa Monica, Calif.: RAND Corporation, RM-4718-PR, 1966. (Declassified by Air Force Declassification Office on October 23, 2018.) As of December 9, 2020:
https://www.rand.org/pubs/research_memoranda/RM4718.html

Hammer, J[ohn] G., and Armas Laupa, *On Hardened Basing of B-52 Aircraft*, Santa Monica, Calif.: RAND Corporation, D-9513, 1961.

Hammer, J[ohn] G., and C[harles] A. Sandoval, *Comparison and Evaluation of Protective Alert Shelter Concepts for SAC Aircraft*, Santa Monica, Calif.: RAND Corporation, D-8740, 1961. As of December 9, 2020:
https://www.rand.org/pubs/documents/D8740.html

Heginbotham, Eric, Michael Nixon, Forrest E. Morgan, Jacob L. Heim, Jeff Hagen, Sheng Li, Jeffrey Engstrom, Martin C. Libicki, Paul DeLuca, David A. Shlapak, David R. Frelinger, Burgess Laird, Kyle Brady, and Lyle J. Morris, *The U.S.-China Military Scorecard: Forces, Geography, and the Evolving Balance of Power, 1996–2017*, Santa Monica, Calif.: RAND Corporation, RR-392-AF, 2015. As of December 9, 2020: https://www.rand.org/pubs/research_reports/RR392.html

Hill, J[erald] E., J[ohn] J. O'Sullivan, and Marc Peter, Jr., *Restoration of Runways Following Attack*, Santa Monica, Calif.: RAND Corporation, D-3573-PR, 1956.

Jackson, Victor G., *Use of Alternative Recovery Bases by Air Defense Interceptor Aircraft*, Santa Monica, Calif.: RAND Corporation, RM-1152, 1953. (Declassified by USAF on February 1, 1961.)

Jaeger, B[ernard] F., and M[arvin] B. Schaffer, *Tentative Thoughts on Non-Nuclear IRBM's for Attacking Parked Aircraft*, Santa Monica, Calif.: RAND Corporation, D(L)-11285-PR, 1963. As of February 10. 2021: https://www.rand.org/pubs/documents/D11285.html

Kozaczka, Felix, *Modeling Nuclear Vulnerability of NATO Tactical Airfields*, Santa Monica, Calif.: RAND Corporation, IN-22741-DDPAE, 1973.

Kozaczka, Felix, *Nuclear Vulnerability of NATO Tactical Airfields: Some Calculations of Nuclear Radiation Exposure*, Santa Monica, Calif.: RAND Corporation, IN-22933-DDPAE, 1974. (Declassified on January 3, 1983.)

Kozaczka, Felix, and J[ohn] K. Seavers, *Examination of Warsaw Pact Airbase Attack Capability Against Unsheltered Aircraft in the Central Region of NATO*, Santa Monica, Calif.: RAND Corporation, IN-21822-PR, 1971. (Declassified on December 31, 1979.)

Krase, W[illiam] H., *Target-Marking Systems for RPVs Used as Designator Vehicles for Airbase Attack*, Santa Monica, Calif.: RAND Corporation, IN-21646-PR, 1971. (Declassified on April 21, 1978.)

Krase, W[illiam] H., *Vehicle Estimates for Target Designation RPVs*, Santa Monica, Calif.: RAND Corporation, IN-21682-PR, 1971. (Declassified on April 21, 1978.)

Lamar, D[onald] L., and D[eane] Oberste-Lehn, *Operation Cliff Dweller: Determination of Site Location for Hardened Aircraft Bases*, Santa Monica, Calif.: RAND Corporation, D-8078, 1960.

Lansdowne, Zachary F., *Computing the Closure Probability for an Airfield Defended with Anti-Tactical Ballistic Missiles*, Santa Monica, Calif.: RAND Corporation, WD-3651-A/ACQ, 1987.

Laupa, Armas, *Time Estimates to Repair Cratered Runways*, Santa Monica, Calif.: RAND Corporation, D-19937-PR, 1970. (Declassified on January 20, 1977.)

Levine, Robert A., *Hard Homes for Heavy-Bomber Crews*, Santa Monica, Calif.: RAND Corporation, D-6744-PR, 1959.

Lewis, Donald E., Bruce W. Don, Robert M. Paulson, and Willis W. Ware, *A Perspective on the USAFE Collocated Operating Base System*, Santa Monica, Calif.: RAND Corporation, N-2366-AF, 1986. As of December 9, 2020:
https://www.rand.org/pubs/notes/N2366.html

Lostumbo, Michael J., David R. Frelinger, James Williams, and Barry Wilson, *Air Defense Options for Taiwan: An Assessment of Relative Costs and Operational Benefits*, Santa Monica, Calif.: RAND Corporation, RR-1051-OSD, 2016. As of December 9, 2020:
https://www.rand.org/pubs/research_reports/RR1051.html

Lostumbo, Michael J., Michael J. McNerney, Eric Peltz, Derek Eaton, David R. Frelinger, Victoria A. Greenfield, John Halliday, Patrick Mills, Bruce R. Nardulli, Stacie L. Pettyjohn, Jerry M. Sollinger, and Stephen M. Worman, *Overseas Basing of U.S. Military Forces: An Assessment of Relative Costs and Strategic Benefits*, Santa Monica, Calif.: RAND Corporation, RR-201-OSD, 2013. As of December 9, 2020:
https://www.rand.org/pubs/research_reports/RR201.html

McGlothlin, W[illiam] H., *A Method for Studying the SAC Base Dispersal Problem*, Santa Monica, Calif.: RAND Corporation, D(L)-4740, 1957.

Miller, Sidney H., *An Approach to Studying Methods of Achieving Air Superiority by Attacking Enemy Airfields*, Santa Monica, Calif: RAND Corporation, 1974, not available to the general public.

Mills, Gary F., and Gerald C. Sumner, *Costs of Airbase Survivability Options*, Santa Monica, Calif.: RAND Corporation, IN-25488-AF, 1988.

Mills, Patrick, James A. Leftwich, John G. Drew, Daniel P. Felten, Josh Girardini, John P. Godges, Michael J. Lostumbo, Anu Narayanan, Kristin Van Abel, Jonathan William Welburn, and Anna Jean Wirth, *Building Agile Combat Support Competencies to Enable Evolving Adaptive Basing Concepts*, Santa Monica, Calif.: RAND Corporation, RR-4200-AF, 2020. As of December 9, 2020:
https://www.rand.org/pubs/research_reports/RR4200.html

Neu, Carl Richard, *Attacking Hardened Air Bases (AHAB): A Decision Analysis Aid for the Tactical Commander*, Santa Monica, Calif.: RAND Corporation, R-1422-PR, 1974. As of December 9, 2020:
https://www.rand.org/pubs/reports/R1422.html

O'Sullivan, J[ohn] J., *Time to Clear Building Debris from Air Base Runways*, Santa Monica, Calif.: RAND Corporation, D(L)-5032, 1958.

Peter, Marc, Jr., *Estimates of Aircraft Shelter Costs*, Santa Monica, Calif.: RAND Corporation D-3347-PR, 1955.

Peter, Marc, Jr., and Charles Sandoval, *Variations in Aircraft Shelter Costs*, Santa Monica, Calif.: RAND Corporation, D-3509-PR, 1956.

Priebe, Miranda, Alan Vick, Jacob Heim, and Meagan Smith, *Distributed Operations in a Contested Environment: Implications for USAF Force Presentation*, Santa Monica, Calif.: RAND Corporation, RR-2959-AF, 2019. As of December 9, 2020: https://www.rand.org/pubs/research_reports/RR2959.html

RAND Corporation, Cost Analysis Section, *The Cost of Decreasing Vulnerability of Air Bases by Dispersal: Dispersing a B-36 Wing*, Santa Monica, Calif., R-235, 1952. As of December 9, 2020: https://www.rand.org/pubs/reports/R235.html

RAND staff, *Special Staff Report: The Selection of Strategic Air Bases*, Santa Monica, Calif.: RAND Corporation, R-244-S, 1953. (Classification canceled by USAF on July 26, 1963. Air Force Declassification Office reviewed and confirmed declassification on August 10, 2015.)

RAND Staff Report, *Protecting U.S. Power to Strike Back in the 1950s and 1960s*, Santa Monica, Calif.: RAND Corporation, R-290, 1956. (Declassified by Air Force Declassification Office on April 18, 2016.)

Sandoval, Charles A., *Size and Cost of Shelter Doors Subjected to Long Duration Nuclear Blast Loading*, Santa Monica, Calif.: RAND Corporation, D-4630, 1957.

Sandoval, C[harles] A., *A Handbook for Estimating Material Requirements and Costs of Shelter Doors Subjected to Long-Duration Blast Loading*, Santa Monica, Calif.: RAND Corporation, RM-2277, 1958. As of December 9, 2020: https://www.rand.org/pubs/research_memoranda/RM2277.html

Sandoval, Charles A., *Report of a Meeting Concerning B-52 Hardening*, Santa Monica, Calif.: RAND Corporation, D-6894-PR, 1959.

Schelling, Thomas C., *Sitting Ducks or Decoys: The High Cost of SAC Dispersal to Large-City Airfields*, Santa Monica, Calif.: RAND Corporation, D-7329, 1960.

Sharkey, E[dward] H., *Some Hand-Done Calculations on Attacks Against Runways*, Santa Monica, Calif.: RAND Corporation, D-18821-PR, 1969.

Sharkey, E[dward] H., *Estimated CEP of Weapon Delivery for Attacks Against Aircraft Shelters*, Santa Monica, Calif.: RAND Corporation, D-20250-PR, 1970. (Declassified on January 20, 1977.)

Shlapak, David A., David T. Orletsky, Toy I. Reid, Murray Scot Tanner, and Barry Wilson, *A Question of Balance: Political Context and Military Aspects of the China-Taiwan Dispute*, Santa Monica, Calif.: RAND Corporation, MG-888-SRF, 2009. As of December 9, 2020: https://www.rand.org/pubs/monographs/MG888.html

Shlapak, David A., David T. Orletsky, and Barry Wilson, *Dire Strait? Military Aspects of the China-Taiwan Confrontation and Options for U.S. Policy*, Santa Monica, Calif.: RAND Corporation, MR-1217-SRF, 2000. As of December 9, 2020: https://www.rand.org/pubs/monograph_reports/MR1217.html

Shlapak, David A., and Alan Vick, *"Check Six Begins on the Ground": Responding to the Evolving Ground Threat to U.S. Air Force Bases*, Santa Monica, Calif.: RAND Corporation, MR-606-AF, 1995. As of December 9, 2020: https://www.rand.org/pubs/monograph_reports/MR606.html

Skogstad, A[nna] L., R[oger] N. Snow, *Preservation of Tactical Air Combat Potential in Western Europe: Allied Force, Base, and Radar Structure*, Santa Monica, Calif.: RAND Corporation, RM-1310, 1954. (Declassified by Air Force Declassification Office on August 13, 1965.)

Snow, R[oger] N., *An Effectiveness Model for Multiple Attacks Against an Airbase Area Complex*, Santa Monica, Calif.: RAND Corporation, R-1639-PR, 1975. As of December 9, 2020: https://www.rand.org/pubs/reports/R1639.html

Stillion, John, and David T. Orletsky, *Airbase Vulnerability to Conventional Cruise-Missile and Ballistic-Missile Attacks: Technology, Scenarios, and U.S. Air Force Responses*, Santa Monica, Calif.: RAND Corporation, MR-1028-AF, 1999. As of December 9, 2020: https://www.rand.org/pubs/monograph_reports/MR1028.html

Stockton, A[l] C., *Preservation of Tactical Air Combat Potential in Western Europe: Underground Hangars for Fighter-Bomber Operations*, Santa Monica, Calif.: RAND Corporation, RM-1230, 1954.

Tamplin, A[rthur] R., *Operation Cliff Dweller: Hardening Bases, Atmospheric Control and Disease*, Santa Monica, Calif.: RAND Corporation, D-8020, 1960.

Tatum, F[reeman] A., and L[ouis] N. Rowell, *PROBE I: A Differential Equation Model for Comparing Fighter Escort and Airbase Attack Systems in a Counter-Air Operation*, Santa Monica, Calif.: RAND Corporation, R-1413-PR, 1974. As of December 9, 2020: https://www.rand.org/pubs/reports/R1413.html

Thomas, Brent, Mahyar A. Amouzegar, Rachel Costello, Robert A. Guffey, Andrew Karode, Christopher Lynch, Kristin F. Lynch, Ken Munson, Chad J. R. Ohlandt, Daniel M. Romano, Ricardo Sanchez, Robert S. Tripp, and Joseph V. Vesely, *Project AIR FORCE Modeling Capabilities for Support of Combat Operations in Denied Environments*, Santa Monica, Calif.: RAND Corporation, RR-427-AF, 2015. As of December 9, 2020:
https://www.rand.org/pubs/research_reports/RR427.html

Tompkins, Thomas C., *An Airbase Ground Attack Scenario in Central Europe*, Santa Monica, Calif.: RAND Corporation, WD-3549-AF, 1987.

Tuck, R[ichard] E., *Preservation of Tactical Air Combat Potential in Western Europe: Guided Missile Defense Potential*, Santa Monica, Calif.: RAND Corporation, RM-1312, 1954. (Declassified by Air Force Declassification Office on April 18, 2016.) As of December 9, 2020:
https://www.rand.org/pubs/research_memoranda/RM1312.html

Vick, Alan J., *Snakes in the Eagle's Nest: A History of Ground Attacks on Air Bases*, Santa Monica, Calif.: RAND Corporation, MR-553-AF, 1995. As of December 9, 2020:
https://www.rand.org/pubs/monograph_reports/MR553.html

Vick, Alan J., *Air Base Attacks and Defensive Counters: Historical Lessons and Future Challenges*, Santa Monica, Calif.: RAND Corporation, RR-968-AF, 2015. As of December 9, 2020:
https://www.rand.org/pubs/research_reports/RR968.html

Vick, Alan, Sean Zeigler, Julia Brackup, and John Speed Meyers, *Air Base Defense: Rethinking Army and Air Force Roles and Functions*, Santa Monica, Calif.: RAND Corporation, RR-4368-AF, 2020. As of December 9, 2020:
https://www.rand.org/pubs/research_reports/RR4368.html

Wakeley, Jay T., *Dispersal: An Expedient for Protecting SAC*, Santa Monica, Calif.: RAND Corporation, S-113, 1958. (Declassified by Air Force Declassification Office on April 19, 2016.)

Ward, G[erri] P., *B-52 Main and Satellite Base Installations Cost: Warm and Cold Climate*, Santa Monica, Calif.: RAND Corporation, D-2908-PR, 1955.

Ward, G[erri] P., *SAC Dispersal: Estimated Installations Costs of Wing, Main and Satellite Bomber Bases, Hard and Soft*, Santa Monica, Calif.: RAND Corporation, D-3005-PR, 1955.

Ward, G[erri] P., *SAC Dispersal: Estimated Costs of Recovery Bases and Home Shelters*, Santa Monica, Calif.: RAND Corporation, D-2966-1-PR, 1955.

Wegner, Louis H., *The Taxiway Repair Schedule Problem: A Heuristic Rule and a Branch-and-Bound Solution*, Santa Monica, Calif.: RAND Corporation, N-1883-AF, 1982. As of

December 9, 2020:
https://www.rand.org/pubs/notes/N1883.html

Weidlinger, Paul, *An Alert Shelter for the B-58*, Santa Monica, Calif.: RAND Corporation, D-6698-PR, 1959.

Wilson, N[atalie] E., and B[ernard] F. Jaeger, *Runbst-Runway Cutting III: Runway Cutting Programs*, Santa Monica, Calif.: RAND Corporation, D-15197-PR, 1966.

Wirth, Anna Jean, Thomas Light, Daniel M. Romano, Shane Tierney, Ronald G. McGarvey, Moon Kim, Michael J. Lostumbo, Amanda Nguyen, Paul Emslie, and John G. Drew, *Evaluating Alternative Maintenance Manpower Force Structure Concepts for the F-35A*, Santa Monica, Calif.: RAND Corporation, RR-4433-AF, 2020. As of December 9, 2020:
https://www.rand.org/pubs/research_reports/RR4433.html

Wohlstetter, Albert, *Economic and Strategic Considerations in Air Base Location: A Preliminary Review*, Santa Monica, Calif.: RAND Corporation, D-1114, 1951. As of December 9, 2020:
https://www.rand.org/pubs/documents/D1114.html

Wohlstetter, Albert, *The Delicate Balance of Terror*, Santa Monica, Calif.: RAND Corporation, P-1472, 1958. As of December 9, 2020:
https://www.rand.org/pubs/papers/P1472.html

Wohlstetter, Albert, Fred Hoffman, *Defending a Strategic Force After 1960*, Santa Monica, Calif.: RAND Corporation, D-2270, 1954. As of December 9, 2020:
https://www.rand.org/pubs/documents/D2270.html

Wohlstetter, Albert, Fred Hoffman, Michael E. Arnsten, *Measures to Protect Airbase Bulk Fuel Stocks*, Santa Monica, Calif.: RAND Corporation, RM-1398, 1954. As of December 9, 2020:
https://www.rand.org/pubs/research_memoranda/RM1398.html

Wohlstetter, Albert, F[red] S. Hoffman, R[obert] J. Lutz, and H[enry] S. Rowen, *Selection and Use of Strategic Air Bases*, Santa Monica, Calif.: RAND Corporation, R-266, 1954. As of December 9, 2020:
https://www.rand.org/pubs/reports/R0266.html

Early RAND Reports on Nuclear Strategy and Deterrence

Brodie, Bernard, *The Anatomy of Deterrence*, Santa Monica, Calif.: RAND Corporation, RM-2218, 1958. As of December 9, 2020:
https://www.rand.org/pubs/research_memoranda/RM2218.html

Brodie, Bernard, *Strategy in the Missile Age*, Santa Monica, Calif.: RAND Corporation, R-335, 1959. (Simultaneously published as a book by Princeton University Press. See the "Books" section for full citation.)

Goldhamer, Herbert, and Andrew W. Marshall, with the assistance of Nathan Leites, *The Deterrence and Strategy of Total War, 1959–1961: A Method of Analysis*, Santa Monica, Calif.: RAND Corporation, RM-2301, 1959. As of February 5, 2020: https://www.rand.org/pubs/research_memoranda/RM2301.html

Kahn, Herman, *Some Specific Suggestions for Achieving Early Non-Military Defense Capabilities and Initiating Long-Range Programs*, Santa Monica, Calif.: RAND Corporation, RM-2206-RC, 1958. As of December 9, 2020: https://www.rand.org/pubs/research_memoranda/RM2206.html

Kahn, Herman, *The Nature and Feasibility of War and Deterrence*, Santa Monica, Calif.: RAND Corporation, P-1888-RC, 1960. As of December 9, 2020: https://www.rand.org/pubs/papers/P1888.html

Schelling, T[homas] C., *The Reciprocal Fear of Surprise Attack*, Santa Monica, Calif.: RAND Corporation, P-1342, 1958. As of December 9, 2020: https://www.rand.org/pubs/papers/P1342.html

Schelling, T[homas] C., *The Threat That Leaves Something to Chance*, Santa Monica, Calif.: RAND Corporation, D(L)-6936, 1959.

Other RAND Reports

Amouzegar, Mahyar A., Ronald G. McGarvey, Robert S. Tripp, Louis Luangkesorn, Thomas Lang, and Charles Robert Roll, Jr., *Evaluation of Options for Overseas Combat Support Basing*, Santa Monica, Calif.: RAND Corporation, MG-421-AF, 2006. As of December 9, 2020: https://www.rand.org/pubs/monographs/MG421.html

Ansoff, H. Igor, W. W. Baldwin, D. J. Davis, Norman Maurice Kaplan, Paul Kecskemeti, and Albert Wohlstetter, *Outline of a Study for the Plans Analysis Section*, Santa Monica, Calif.: RAND Corporation, D-937, 1951. As of December 9, 2020: https://www.rand.org/pubs/documents/D937.html

Bitzinger, Richard A., *Assessing the Conventional Balance in Europe, 1945–1975*, Santa Monica, Calif.: RAND Corporation, N-2859-FF/RC, May 1989. As of December 9, 2020: https://www.rand.org/pubs/notes/N2859.html

Bowie, Christopher, Fred Frostic, Kevin Lewis, John Lund, David Ochmanek, and Phillip Propper, *The New Calculus: Analyzing Airpower's Changing Role in Joint Theater*

Campaigns, Santa Monica, Calif.: RAND Corporation, MR-149-AF, 1993. As of December 9, 2020:
https://www.rand.org/pubs/monograph_reports/MR149.html

Canby, Steven L., *NATO Military Policy: Obtaining Conventional Comparability with the Warsaw Pact*, Santa Monica, Calif: RAND Corporation, R-1088-ARPA, 1973. As of December 9, 2020:
https://www.rand.org/pubs/reports/R1088.html

Cliff, Roger, Mark Burles, Michael S. Chase, Derek Eaton, and Kevin L. Pollpeter, *Entering the Dragon's Lair: Chinese Antiaccess Strategies and Their Implications for the United States*, Santa Monica, Calif.: RAND Corporation, MG-524-AF, 2007. As of December 9, 2020:
https://www.rand.org/pubs/monographs/MG524.html

Cliff, Roger, John F. Fei, Jeff Hagen, Elizabeth Hague, Eric Heginbotham, and John Stillion, *Shaking the Heavens and Splitting the Earth: Chinese Air Force Employment Concepts in the 21st Century*, Santa Monica, Calif.: RAND Corporation, MG-915-AF, 2011. As of December 9, 2020:
https://www.rand.org/pubs/monographs/MG915.html

Davies, Merton E., and William R. Harris, *RAND's Role in the Evolution of Balloon and Satellite Observation Systems and Related U.S. Space Technology*, Santa Monica, Calif.: RAND Corporation, R-3692-RC, 1988. As of December 9, 2020:
https://www.rand.org/pubs/reports/R3692.html

Davis, Paul K., "Analytic Methods," in RAND Corporation, ed., *Project AIR FORCE: 1946–1996*, Santa Monica, Calif.: RAND Corporation, 1996, pp. 47–52. As of December 19, 2020:
https://www.rand.org/content/dam/rand/www/external/publications/PAFbook.pdf

Davis, Paul K., *Analysis to Inform Defense Planning Despite Austerity*, Santa Monica, Calif.: RAND Corporation, RR-482-OSD, 2014. As of December 9, 2020:
https://www.rand.org/pubs/research_reports/RR482.html

Elliott, Mai, *RAND in Southeast Asia: A History of the Vietnam War Era*, Santa Monica, California: RAND Corporation, CP-564-RC, 2010. As of December 9, 2020:
https://www.rand.org/pubs/corporate_pubs/CP564.html

Enthoven, Alain C., and K. Wayne Smith, *How Much Is Enough? Shaping the Defense Program 1961–1969*, Santa Monica, Calif.: RAND Corporation, CB-403, 2005. As of December 14, 2020:
https://www.rand.org/pubs/commercial_books/CB403.html

Fisher, Gene H., and Warren E. Walker, *Operations Research and the RAND Corporation*, Santa Monica, Calif.: RAND Corporation, P-7857, 1994. As of December 9, 2020:
https://www.rand.org/pubs/papers/P7857.html

Gruenberger, F[red] J[oseph], *The History of the JOHNNIAC*, Santa Monica, Calif.: RAND Corporation, RM-5654-PR, 1968. As of December 9, 2020:
https://www.rand.org/pubs/research_memoranda/RM5654.html

Hamilton, Thomas, and David Ochmanek, *Operating Low-Cost, Reusable, Unmanned Aerial Vehicles in Contested Environments: Preliminary Evaluation of Operational Concepts*, Santa Monica, Calif.: RAND Corporation, RR-4407-AF, 2020. As of December 9, 2020:
https://www.rand.org/pubs/research_reports/RR4407.html

Hoehn, Andrew R., Adam Grissom, David A. Ochmanek, David A. Shlapak, and Alan J. Vick, *A New Division of Labor: Meeting America's Security Challenges Beyond Iraq*, Santa Monica, Calif: RAND Corporation, MG-499-AF, 2007. As of December 9, 2020:
https://www.rand.org/pubs/monographs/MG499.html

Khalilzad, Zalmay M., and David Ochmanek, eds., *Strategic Appraisal 1997*: *Strategy and Defense Planning for the 21st Century*, Santa Monica, Calif.: RAND Corporation, MR-826-AF, 1997. As of December 9, 2020:
https://www.rand.org/pubs/monograph_reports/MR826.html

Killingsworth, Paul, Lionel A. Galway, Eiichi Kamiya, Brian Nichiporuk, Robert S. Tripp, and James C. Wendt, *Flexbasing: Achieving Global Presence for Expeditionary Aerospace Forces*, Santa Monica, Calif.: RAND Corporation, MR-1113-AF, 2000. As of December 9, 2020:
https://www.rand.org/pubs/monograph_reports/MR1113.html

Kugler, Richard L., *The Great Strategy Debate: NATO's Evolution in the 1960s*, Santa Monica, Calif.: RAND Corporation, N-3252-FF/RC, 1991. As of December 9, 2020:
https://www.rand.org/pubs/notes/N3252.html

Legge, J. Michael, *Theater Nuclear Weapons and the NATO Strategy of Flexible Response*, Santa Monica, Calif: RAND Corporation, R-2964-FF, 1983. As of December 9, 2020:
https://www.rand.org/pubs/reports/R2964.html

Leites, Nathan, and Charles Wolf, Jr., *Rebellion and Authority: An Analytic Essay on Insurgent Conflicts*, Santa Monica, Calif.: RAND Corporation, R-462-ARPA, 1970. As of December 9, 2020:
https://www.rand.org/pubs/reports/R0462.html

Long, Austin, *On "Other War:" Lessons from Five Decades of RAND Counterinsurgency Research*, Santa Monica, Calif.: RAND Corporation, MG-482-OSD, 2006. As of December 9, 2020:
https://www.rand.org/pubs/monographs/MG482.html

Long, Austin, *Deterrence—From Cold War to Long War: Lessons from Six Decades of RAND Research*, Santa Monica, Calif.: RAND Corporation, MG-636-OSD/AF, 2008. As of

December 9, 2020:
https://www.rand.org/pubs/monographs/MG636.html

Marks, Shirley L., *The JOSS Years: Reflections on an Experiment*, Santa Monica, Calif: RAND
Corporation, R-918, December 1971. As of December 9, 2020:
https://www.rand.org/pubs/reports/R0918.html

Mills, Patrick, John G. Drew, John A. Ausink, Daniel M. Romano, and Rachel Costello,
Balancing Agile Combat Support Manpower to Better Meet the Future Security Environment,
Santa Monica, Calif: RAND Corporation, RR-337-AF, 2014. As of December 9, 2020:
https://www.rand.org/pubs/research_reports/RR337.html

Mueller, Karl P., ed., *Precision and Purpose: Airpower in the Libyan Civil War*, Santa Monica,
Calif.: RAND Corporation, RR-676-AF, 2015. As of December 9, 2020:
https://www.rand.org/pubs/research_reports/RR676.html

Quade, E. S., *Military Systems Analysis*, Santa Monica, Calif.: RAND Corporation, RM-3452-
PR, 1963. As of December 9, 2020:
https://www.rand.org/pubs/research_memoranda/RM3452.html

RAND Corporation, "A Brief History of RAND," webpage, undated. As of December 9, 2020:
https://www.rand.org/about/history/a-brief-history-of-rand.html

RAND Corporation, *Preliminary Design of an Experimental World-Circling Spaceship*, Santa
Monica, Calif.: RAND Corporation, SM-11827, 1946. As of December 9, 2020:
https://www.rand.org/pubs/special_memoranda/SM11827.html

RAND Corporation, *Fostering Innovation in the Defense Department: Examples from RAND's
Federally Funded Research and Development Centers*, Santa Monica, Calif.: RAND
Corporation, CP-852, June 2016. As of December 9, 2020:
https://www.rand.org/pubs/corporate_pubs/CP852.html

Rich, Michael D., *RAND's Role in the CORONA Program: Remarks on the 35th Anniversary of
the First Successful Mission*, Santa Monica, Calif.: RAND Corporation, P-8017, 1998. As of
December 9, 2020:
https://www.rand.org/pubs/papers/P8017.html

Shishko, Robert, *The European Conventional Balance: A Primer*, Santa Monica, Calif.: RAND
Corporation, P-6707, 1981. As of December 9, 2020:
https://www.rand.org/pubs/papers/P6707.html

Snyder, Don, and Patrick Mills, *Supporting Air and Space Expeditionary Forces: A Methodology
for Determining Air Force Deployment Requirements*, Santa Monica, Calif.: RAND
Corporation, MG-176-AF, 2004. As of December 9, 2020:
https://www.rand.org/pubs/monographs/MG176.html

Stockfisch, J., *Models, Data, and War: A Critique of the Study of Conventional Forces*, Santa Monica, Calif.: RAND Corporation, R-1526-PR, 1975. As of December 9, 2020:
https://www.rand.org/pubs/reports/R1526.html

Thornhill, Paula G., *"Over Not Through": The Search for a Strong, Unified Culture for America's Airmen*, Santa Monica, Calif.: RAND Corporation, OP-386-AF, 2012. As of December 9, 2020:
https://www.rand.org/pubs/occasional_papers/OP386.html

Tripp, Robert S., *The Line Between Disorder and Order: Reflections on RAND's Role in the Evolution of Air Force Logistics Thought and Practice*, Santa Monica, Calif.: RAND Corporation, RR-3131-AF, 2020. As of December 9, 2020:
https://www.rand.org/pubs/research_reports/RR3131.html

Tripp, Robert S., Lionel Galway, Paul S. Killingsworth, Eric Peltz, Timothy L. Ramey, and John G. Drew, *Supporting Expeditionary Aerospace Forces: An Integrated Agile Combat Support Planning Framework*, Santa Monica, Calif.: RAND Corporation, MR-1056-AF, 1999. As of December 9, 2020:
https://www.rand.org/pubs/monograph_reports/MR1056.html

Vick, Alan J., *Proclaiming Airpower: Air Force Narratives and American Public Opinion from 1917 to 2014*, Santa Monica, Calif.: RAND Corporation, RR-1044-AF, 2015. As of December 9, 2020:
https://www.rand.org/pubs/research_reports/RR1044.html

Vick, Alan J., *Force Presentation in U.S. Air Force History and Airpower Narratives*, Santa Monica, Calif.: RAND Corporation, RR-2363-AF, 2018. As of December 9, 2020:
https://www.rand.org/pubs/research_reports/RR2363.html

Ware, Willis H., *RAND and the Information Evolution: A History in Essays and Vignettes*, Santa Monica, Calif.: RAND Corporation, CP-537-RC, 2008. As of December 9, 2020:
https://www.rand.org/pubs/corporate_pubs/CP537.html

Wohlstetter, Albert, *Systems Analysis Versus Systems Design*, Santa Monica, Calif.: RAND Corporation, P-1530, 1958. As of December 9, 2020:
https://www.rand.org/pubs/papers/P1530.html

Wohlstetter, Albert, *Theory and Opposed-Systems Design*, Santa Monica, Calif.: RAND Corporation, D-16001-1, 1968. As of December 9, 2020:
https://www.rand.org/pubs/documents/D16001-1.html

Government Documents

Barclay, Nadine Y., "RPA Prophecy Fulfilled, Oldest RPA Squadron Celebrates 20 Years," Air Combat Command website, July 29, 2015. As of April 10, 2010: https://www.acc.af.mil/News/Article-Display/Article/660480/rpa-prophecy-fulfilled-oldest-rpa-squadron-celebrates-20-years/

Benson, Lawrence R., *USAF Aircraft Basing in Europe, North Africa, and the Middle East, 1945–1980*, Ramstein Air Base, Germany: Headquarters, U.S. Air Forces in Europe, 1981. (Declassified in 2011 by the Air Force History Office.)

Bradley, Emily A., "Andersen Airmen Learn Innovative Airfield Damage Repair Capability," U.S. Air Force, February 28, 2014. As of December 9, 2020: http://www.af.mil/News/ArticleDisplay/tabid/223/Article/473449/andersen-airmen-learn-innovative-airfield-damage-repair-capability.aspx

Davis, Richard G., *Anatomy of a Reform: The Expeditionary Aerospace Force*, Washington, D.C.: Air Force History and Museums Program, 2003.

Director of Central Intelligence, *Warsaw Pact Nonnuclear Threat to NATO Airbases in Central Europe: National Intelligence Estimate*, Washington, D.C.: Central Intelligence Agency, NIE 11/20-6-84, October 25, 1984; declassified; referenced June 9, 2014. As of December 19, 2020: http://www.foia.cia.gov/sites/default/files/document_conversions/89801/DOC_0000278545.pdf

Finletter, Thomas Knight, *Survival in the Air Age: A Report*, Washington, D.C.: President's Air Policy Commission, 1948.

Glasstone, Samuel, and Philip J. Dolan, *The Effects of Nuclear Weapons*, Washington, D.C.: U.S. Departments of Defense and Energy, 1977.

Johnson, Dani, "Kadena Prepares for Typhoon," July 12, 2007, U.S. Air Force, July 12, 2007. As of December 9, 2020: https://www.af.mil/News/Article-Display/Article/126308/kadena-prepares-for-typhoon/

Kaplan, Lawrence S., Ronald D. Landa, and Edward J. Drea, *History of the Office of the Secretary of Defense, Volume V, The McNamara Ascendancy: 1961–1965*, Washington, D.C.: Historical Office, Office of the Secretary of Defense, 2006.

Larm, Dennis, *Expendable Remotely Piloted Vehicles for Strategic Offensive Airpower Roles*, Maxwell Air Force, Ala.: Air University Press, June 1996. As of April 10, 2020: https://media.defense.gov/2017/Dec/27/2001861507/-1/-1/0/T_0028_LARM_EXPENDABLE_REMOTELY_PILOTED.PDF

Leighton, Richard M., *History of the Office of the Secretary of Defense, Volume III: Strategy, Money, and the New Look, 1953–1956*, Washington, D.C.: Historical Office, Office of the Secretary of Defense, 2001.

Mattis, Jim, *Summary of the 2018 National Defense Strategy of The United States of America: Sharpening the American Military's Competitive Edge*, Washington, D.C.: U.S. Department of Defense, 2018.

McNamara, Robert S., *Statement on the Fiscal Year 1969–73 Defense Program and the 1969 Defense Budget*, Washington, D.C.: U.S. Department of Defense, 1968.

Meilinger, Phillip S., *Bomber: The Formation and Early Years of Strategic Air Command*, Maxwell Air Force Base, Alabama: Air University Press, 2012.

Moody, Walton S., Jacob Neufeld, and R. Cargill Hall, "The Emergence of the Strategic Air Command," in Bernard Nalty, *Winged Shield, Winged Sword: A History of the United States Air Force, Volume II, 1950–1997*, Washington, D.C.: Air Force History and Museums Program, 1997.

Narducci, Henry, *Strategic Air Command and the Alert Program: A Brief History*, Offutt Air Force Base, Nebraska: Office of the Historian, Headquarters, Strategic Air Command, 1988.

Office of the Federal Register, National Archives and Records Administration, "News Conference of Secretary of Defense Melvin R. Laird Following the President's Announcement," January 13, 1972, in *Weekly Compilation of Presidential Documents*, Quarterly Index, First Quarter, January–March 1972.

Romjue, John L., *From Active Defense to AirLand Battle: The Development of Army Doctrine 1973–1982*, Fort Leavenworth, Kans.: U.S. Army Training and Doctrine Command, TRADOC Historical Monograph Series, June 1984.

Rostker, Bernard, *Information Paper: Iraq's SCUD Ballistic Missiles*, Washington, D.C.: U.S. Department of Defense, interim paper, July 25, 2000.

Stokes, Mark A., *China's Strategic Modernization: Implications for the United States*, Carlisle, Pa.: Strategic Studies Institute, U.S. Army War College, September 1999. As of April 11, 2020:
http://purl.access.gpo.gov/GPO/LPS12109

Strategic Air Command, Office of the Historian, *Alert Operations and the Strategic Air Command: 1957–1991*, Offutt Air Force Base, Neb., 1991.

U.S. Air Force, *Force Protection*, Air Force Doctrine Document 2-4.1, Maxwell Air Force Base, Ala.: Air Force Doctrine Center, October 29, 1999. As of April 15, 2020:
https://www.globalsecurity.org/military/library/policy/usaf/afdd/2-4-1/afdd2-4-1.pdf

U.S. Department of Energy, Nevada Operations Office, *United States Nuclear Tests: July 1945 Through September 1992*, Las Vegas, Nev., 2000.

U.S. Department of State, Treaty Between the United States of America and The Union of Soviet Socialist Republics on The Limitation of Anti-Ballistic Missile Systems (ABM Treaty), signed May 26, 1972. As of April 10, 2020: https://2009-2017.state.gov/t/avc/trty/101888.htm

U.S. National Air and Space Intelligence Center, *Ballistic and Cruise Missile Threat*, Wright-Patterson Air Force Base, Ohio, 2013. As of December 9, 2020: https://apps.dtic.mil/dtic/tr/fulltext/u2/a582843.pdf

U.S. Senate, "Study of Airpower," hearings before the Subcommittee of the Air Force of the Committee on Armed Services, Part XXI, 84th Congress, 2nd Session, June 26 and 28, 1956.

Weitze, Karen, *Eglin Air Force Base, 1931–1991: Installation Buildup for Research, Test, Evaluation and Training*, Eglin Air Force Base, Fla.: Air Force Materiel Command, 2001.

Books

Ball, Desmond, *Politics and Force Levels: The Strategic Missile Program of the Kennedy Administration*, Berkeley, Calif.: University of California Press, 1980.

Berhow, Mark, *U.S. Strategic and Defensive Missile Systems, 1950–2004*, Oxford, UK: Osprey Publishing, 2005.

Bowie, Robert R., and Richard H. Immerman, *Waging Peace: How Eisenhower Shaped an Enduring Cold War Strategy*, Oxford, UK: Oxford University Press, 1998.

Brodie, Bernard, ed., *The Absolute Weapon: Atomic Power and World Order*, New York: Harcourt and Brace, 1946.

Brodie, Bernard, *Strategy in the Missile Age*, Princeton, N.J.: Princeton University Press, 1959.

Bruce-Briggs, Barry, *Supergenius: The Mega-Worlds of Herman Kahn*, Morrisville, N.C.: Lulu Enterprises, 2005.

Call, Steve, *Selling Air Power: Military Aviation and American Popular Culture After World War II*, College Station, Tex.: Texas A&M Press, 2009.

Castle, Ian, *The Zeppelin Base Raids: Germany 1914*, Oxford, UK: Osprey Publishing, 2011.

Caudill, Shannon, ed., *Defending Air Bases in An Age of Insurgency*, Maxwell Air Force Base, Ala.: Air University Press, 2014.

Caudill, Shannon, ed., *Defending Air Bases In An Age of Insurgency: Volume II*, Maxwell Air Force Base, Ala.: Air University Press, 2019.

De Longe, Merrill E., *Modern Airfield Planning and Concealment*, New York: Chicago Pitman Publishing Corporation, 1943.

Douhet, Giulio, *The Command of the Air*, trans. Dino Ferrari, Washington, D.C.: Office of Air Force History Imprint, 1983 (originally published in Italian in 1921).

Ellsberg, Daniel, *The Doomsday Machine: Confessions of a Nuclear War Planner*, New York: Bloomsbury Publishing, 2017.

Fox, Roger, *Air Base Defense in the Republic of Vietnam, 1961–1973*, Washington, D.C.: Office of Air Force History, 1979.

Franks, Norman L. R., *Battle of the Airfields: Operation Bodenplatte, 1 January, 1945*, London: Grub Street, 1994.

Fukuyama, Francis, *The End of History and the Last Man*, New York: Free Press, 1992.

Ghamari-Tabrizi, Sharon, *The Worlds of Herman Kahn: The Intuitive Science of Thermonuclear War*, Cambridge, Mass.: Harvard University Press, 2005.

Hallion, Richard P., *Storm Over Iraq: Air Power and the Gulf War*, Washington, D.C.: Smithsonian Institution Press, 1992.

Holloway, David, *Stalin and the Bomb: The Soviet Union and Atomic Energy, 1939–1956*, New Haven, Conn.: Yale University Press, 1994.

Hook, Steven W., and John W. Spanier, *American Foreign Policy Since World War II*, Thousand Oaks, Calif.: CQ Press, 2018.

Hopkins, J. C., and Sheldon A. Goldberg, *The Development of Strategic Air Command: 1946–1986 (The Fortieth Anniversary History)*, Offutt Air Force Base, Neb.: Office of the Historian, Headquarters Strategic Air Command, 1986.

Kahn, Herman, *On Thermonuclear War*, Princeton, N.J.: Princeton University Press, 1960.

Kaplan, Fred, *The Wizards of Armageddon*, New York: Simon and Schuster, 1983.

Kozak, Warren, *LeMay: The Life and Wars of General Curtis LeMay*, Washington, D.C.: Regnery History, 2009.

Kreis, John F., *Air Warfare and Air Base Air Defense, 1914–1973*, Washington, D.C.: Office of Air Force History, 1988.

Krepinevich, Andrew F., and Barry D. Watts, *The Last Warrior: Andrew Marshall and the Shaping of Modern American Defense Strategy*, New York: Basic Books, 2015.

Lambeth, Benjamin S., *The Transformation of American Air Power*, Ithaca, N.Y.: Cornell University Press, 2000.

Lambeth, Benjamin S., *The Unseen War: Allied Air Power and the Takedown of Saddam Hussein*, Annapolis, Md.: Naval Institute Press, 2013.

Leffler, Melvyn P., *A Preponderance of Power: National Security, the Truman Administration, and the Cold War*, Palo Alto, Calif.: Stanford University Press, 1992.

Leffler, Melvyn P., *For the Soul of Mankind: The United States, the Soviet Union, and the Cold War*, New York: Hill and Wang, 2007.

LeMay, Curtis E., *America Is in Danger*, New York: Funk and Wagnalls, 1968.

LeMay, Curtis E., with MacKinlay Kantor, *Mission with LeMay: My Story*, Garden City, N.Y.: Doubleday, 1965.

McMahon, Robert J., *The Cold War in the Third World*, Oxford, UK: Oxford University Press, 2013.

Mearsheimer, John J., *Conventional Deterrence*, Ithaca, N.Y.: Cornell University Press, 1983.

Meilinger, Phillip S., *Hoyt S. Vandenberg: The Life of a General*, Bloomington, Ind.: Indiana University Press, 1989.

Meilinger, Phillip S., *10 Propositions Regarding Air Power*, Washington, D.C.: Air Force History and Museums Program, 1995.

Meilinger, Phillip S., ed., *The Paths of Heaven: The Evolution of Airpower Theory*, Maxwell Air Force Base, Ala.: Air University Press, 1997.

Morgan, Mark L., and Mark A. Berhow, *Rings of Supersonic Steel: Air Defenses of the United States Army 1950–1979: An Introductory History and Site Guide*, Bodega Bay, Calif.: Hole in the Head Press, 2010.

Nalty, Bernard C., ed., *Winged Shield, Winged Sword: A History of the United States Air Force*, Volume II, *1950–1997*, Washington, D.C.: U.S. Air Force, 1997.

Neufeld, Jacob, *The Development of Ballistic Missiles in the United States Air Force, 1945–1960*, Washington, D.C.: Office of Air Force History, 1990.

O'Neill, Mark, "The Soviet Air Force, 1917–1991," in Robin Higham, Frederick W. Kagan, eds., *The Military History of the Soviet Union*, New York: Palgrave Macmillan, 2002.

Peebles, Curtis, *High Frontier: The United States Air Force and the Military Space Program*, Washington, D.C.: Air Force History and Museums Program, 1997.

Preble, Christopher A., *John F. Kennedy and the Missile Gap*, DeKalb, Ill.: Northern Illinois University Press, 2004.

Sagan, Scott D., *Moving Targets: Nuclear Strategy and National Security*, Princeton, N.J.: Princeton University Press, 1990.

Schaffel, Kenneth, *The Emerging Shield: The Air Force and the Evolution of Continental Air Defense 1945–1960*, Washington, D.C.: Office of Air Force History, 1991.

Schelling, Thomas C., *The Strategy of Conflict*, Cambridge, Mass.: Harvard University Press, 1960.

Schlesinger, Arthur M., Jr., *A Thousand Days: John F. Kennedy in the White House*, New York: Mariner Books, 2002.

Sherry, Michael S., *The Rise of American Air Power: The Creation of Armageddon*, New Haven, Conn.: Yale University Press, 1987.

Sidoti, S. J., *Airbase Operability: A Study in Airbase Survivability and Post-Attack Recovery*, Canberra: Aerospace Centre, 2001.

Sutton, Boyd, John R. Landry, Malcolm B. Armstrong, Howell M. Estes, and Wesley K. Clark, "Strategic and Doctrinal Implications of Deep Attack Concepts for the Defense of Central Europe," in Keith A. Dunn and William O. Staudenmaier, eds., *Military Strategy in Transition: Defense and Deterrence in the 1980s*, Carlisle Barracks, Pa.: U.S. Army War College, 1984.

Thobo-Carlsen, Paul M., "A Canadian Perspective on Air Base Ground Defense: Ad Hoc Is Not Good Enough," in Shannon W. Caudill, ed., *Defending Air Bases in an Age of Insurgency*, Maxwell Air Force Base, Ala.: Air University Press, 2014.

Tomes, Robert R., *U.S. Defense Strategy from Vietnam to Operation Iraqi Freedom: Military Innovation and the New American Way of War, 1973–2003*, New York: Routlege, 2007.

Westad, Odd, *The Global Cold War: Third World Interventions and the Making of Our Times*, Cambridge, UK: Cambridge University Press, 2007.

Whiting, Kenneth R., *Soviet Air Power*, New York: Routledge, 2019.

Monographs

Bowie, Christopher J., *The Anti-Access Threat and Theater Air Bases*, Washington, D.C.: Center for Strategic and Budgetary Assessments, 2002.

Digby, James, *Precision-Guided Munitions, Adelphi Paper 118*, London: International Institute for Strategic Studies, 1975.

Mako, William, *U.S. Ground Forces and the Defense of Central Europe*, Washington, D.C.: Brookings Institution, 1983.

Rehberg, Carl, and Mark Gunzinger, *Air and Missile Defense at a Crossroads: New Concepts and Technologies to Defend America's Overseas Bases*, Washington, D.C.: Center for Strategic and Budgetary Assessments, 2018.

Periodicals

Alsop, Joseph, "True Missile Gap Picture Belies Pentagon Response," *Eugene Register-Guard*, October 13, 1959.

Apple, R. W., Jr., "War in the Gulf: Scud Attack; Scud Missile Hits a U.S. Barracks, Killing 27," *New York Times*, February 26, 1991. As of April 11, 2020: https://www.nytimes.com/1991/02/26/world/war-in-the-gulf-scud-attack-scud-missile-hits-a-us-barracks-killing-27.html

Arnold, Henry H., "Air Power for Peace," *National Geographic Magazine*, February 1946, pp. 135–193.

Barbara James C., and Robert F. Brown, "Deep Thrust on the Extended Battlefield," *Military Review*, October 1982, pp. 22–32.

Barnes, Julian, "Andrew Marshall, Pentagon's Threat Expert, Dies at 97," *New York Times*, March 26, 2019. As of February 11, 2020: https://www.nytimes.com/2019/03/26/us/politics/andrew-marshall-dead.html

Bell, Raymond E., Jr., "To Protect an Air Base," *Air Power Journal*, Fall 1989, pp. 4–19.

Bingham, Price, "Fighting from the Air Base," *Air Power Journal*, Summer 1987, pp. 32–41.

Bowie, Christopher, "The Lessons of Salty Demo," *Air Force Magazine*, March 2009, pp. 54–57.

Christensen, Thomas, "Posing Problems Without Catching Up: China's Rise and Challenges for U.S. Security Policy," *International Security*, Vol. 25, No. 4, Spring 2001, pp. 5–40.

Coffey, J. I., "The Anti-Ballistic Missile Debate," *Foreign Affairs*, Vol. 45, No. 3, April 1967, pp. 403–413.

Digby, James, "Operations Research and Systems Analysis at RAND, 1948–1967," *OR/MSToday*, Vol. 15, No. 5, December 1989, pp. 10–13.

Flax, Alexander, "Ballistic Missile Defense: Concepts and History," *Daedalus*, Vol. 114, No. 2, Weapons in Space, Vol. I: Concepts and Technologies, Spring 1985, pp. 33–52.

Gibbons-Neff, Thomas, Eric Schmitt, Charlie Savage, and Helene Cooper, "Chaos as Militants Overran Airfield, Killing 3 Americans in Kenya," *New York Times*, January 22, 2020. As of January 31, 2020: https://www.nytimes.com/2020/01/22/world/africa/shabab-kenya-terrorism.html

Gray, Colin S., "The Blitzkrieg: A Premature Burial?" *Military Review*, October 1976, pp. 15–18.

Hampson, Fen Osler, "Groping for Technical Panaceas: The European Conventional Balance and Nuclear Stability," *International Security*, Winter 1983/84, pp. 57–82.

Heim, Jacob L., "The Iranian Missile Threat to Air Bases: A Distant Second to China's Conventional Deterrent," *Air and Space Power Journal*, Vol. 29, No. 4, July–August 2015, pp. 27–50.

Huntington, Samuel P., "Conventional Deterrence and Conventional Retaliation in Europe," *International Security*, Vol. 8, No. 3, Winter 1983–1984, pp. 32–56.

Johnson, Dani, "Kadena Prepares for Typhoon," U.S. Air Force website, July 12, 2007. As of October 20, 2020:
https://www.af.mil/News/Article-Display/Article/126308/kadena-prepares-for-typhoon/

Karber, Phillip A., "In Defense of Forward Defense," *Armed Forces Journal*, May 1984.

Licklider, Roy E., "The Missile Gap Controversy," *Political Science Quarterly*, Vol. 85, No. 4, December 1970.

Mearsheimer, John J., "Why the Soviets Can't Win Quickly in Central Europe," *International Security*, Vol. 7, No. 1, Summer 1982, pp. 3–39.

Mearsheimer, John J., "Nuclear Weapons and Deterrence in Europe," *International Security*, Vol. 9, No. 3, Winter 1984–1985, pp. 19–46.

Menand, Louis, "Fat Man: Herman Kahn and the Nuclear Age," *The New Yorker*, June 20, 2005. As of February 5, 2020:
https://www.newyorker.com/magazine/2005/06/27/fat-man

Preble, Christopher A., "Who Ever Believed in the 'Missile Gap'? John F. Kennedy and the Politics of National Security," *Presidential Studies Quarterly*, Vol. 33, No. 4, December 2003, pp. 801–826.

Record, Jeffrey, "The October War: Burying the Blitzkrieg," *Military Review*, April 1976, pp. 19–21.

Romjue, John "The Evolution of the AirLand Battle Concept," *Air University Review*, May/June 1984.

Rosenberg, David Alan, "The Origins of Overkill: Nuclear Weapons and American Strategy, 1945–1960," *International Security*, Vol. 7, No. 4, Spring 1983, pp. 3–71.

Rubin, Alissa J., "Audacious Raid on NATO Base Shows Taliban's Reach," *New York Times*, September 16, 2012. As of April 11, 2020:
https://www.nytimes.com/2012/09/17/world/asia/green-on-blue-attacks-in-afghanistan-continue.html

Sagan, Scott D., "SIOP-62: The Nuclear War Plan Briefing to President Kennedy," *International Security*, Vol. 12, No. 1, Summer 1987, pp. 22–51.

Starry, Donn A., "Extending the Battlefield," *Military Review*, March 1981, pp. 31–50.

Stillion, John, "Fighting Under Missile Attack," *Air Force Magazine*, August 2009, pp. 34–37.

Spaatz, Carl A., "Air Power in the Atomic Age," *Collier's*, December 8, 1945.

Spaatz, Carl, "Strategic Air Power: Fulfillment of a Concept," *Foreign Affairs*, Vol. 24, No. 3, April 1946, pp. 385–396.

Trachtenberg, Marc, "A 'Wasting Asset': American Strategy and the Shifting Nuclear Balance, 1949–1954," *International Security*, Vol. 13, No. 3, Winter 1988–1989, pp. 5–49.

Vandenberg, Hoyt S., "The Truth About Our Air Power," *Saturday Evening Post*, February 17, 1951.

Weinberger, Sharon, "The Return of the Pentagon's Yoda," *Foreign Policy*, September 12, 2018. As of February 11, 2020:
https://foreignpolicy.com/2018/09/12/the-return-of-the-pentagons-yoda-andrew-marshall/

Weisgerber, Marcus, "Pentagon Debates Policy to Strengthen, Disperse Bases," *Defense News*, April 13, 2014.

Wohlstetter, Albert, "The Delicate Balance of Terror," *Foreign Affairs*, January 1959. As of February 5, 2020:
https://www.foreignaffairs.com/articles/1959-01-01/delicate-balance-terror

Woodmansee, John, "Blitzkrieg and the AirLand Battle," *Military Review*, August 1984.